©
Jill
Moe

## *About the Author*

JOHN MOE is a senior reporter for the nationwide public radio program *Weekend America*. He's a commentator for NPR's *All Things Considered* and writes for the award-winning humor website McSweeneys.net. He lives with his wife and children in Seattle.

# CONSERVATIZE ME

☞ A Lifelong **Lefty's**

**Attempt** to Love

God, Guns, Reagan

&

**Toby Keith** ☜

# john moe

HARPER ● PERENNIAL

NEW YORK ● LONDON ● TORONTO ● SYDNEY

**FOR JILL,**

**with all my gratitude,**

**admiration,**

**and love**

HARPER ● PERENNIAL

A hardcover edition of this book was published in 2006 by William Morrow, an imprint of HarperCollins Publishers.

HarperCollins books may be purchased for educational, business, or sales promotional use. For information please write: Special Markets Department, HarperCollins Publishers, 10 East 53rd Street, New York, NY 10022.

FIRST HARPER PERENNIAL EDITION PUBLISHED 2008.

*Designed by Susan Yang*

Library of Congress Cataloging-in-Publication Data has been applied for.

ISBN: 978-0-06-085402-7 (pbk.)

08 09 10 11 12 NMSG/RRD 10 9 8 7 6 5 4 3 2 1

# contents

# My Mission

## Should I Choose

## to Accept It

☞

**In which the author attempts to tap the
inclinations that could drive him toward
a radical ideological realignment.**

"How do you normally part your hair?" asked Julie, my barber. "To the left or to the right?"

"To the—well, let me see—I guess I never thought about it.
I go like this," I said, smooshing the thinning crop to one side
in a halfhearted motion like I usually do in the morning before
leaving for work. I was a little confused by the mirror but after
quick calculation was able to say, "So I guess to the left."

"No," she said, "your hair goes to the right. You should
comb it that way. You naturally go to the right." She had no
idea how chilling that was for me to hear or why I sat in silent
stricken terror for the rest of the haircut. "Is everything okay?"
she asked, noticing that I was frowning gravely at myself in the
mirror. I told her the haircut was fine. It's me that I was wondering about.

It was mere days before I was to begin a potentially life-
altering experience. I was going to try to make my politics like
my hair, moving from left to right.

I live in Seattle. Republicans still run for office once in
a while around here but it's more of a hobby for them. In

the Seventh Congressional District, which includes most of Seattle, Jim McDermott has been elected to the U.S. House of Representatives in nine straight elections. He cruises to easy victories every time. In the run-up to the invasion of Iraq in 2003, McDermott went to Baghdad along with thirteen-term Michigan representative David Bonior and the two of them were shown around town by emissaries of Saddam Hussein. Ultimately they announced that as far as they could tell, there were no weapons of mass destruction in Iraq. McDermott was heavily criticized for the trip. Conservative columnist George Will said, "McDermott and Bonior are two specimens of what Lenin, referring to Westerners who denied the existence of Lenin's police-state terror, called 'useful idiots.' " The trip took place just a few weeks before the 2002 elections and McDermott, despite being denounced as a traitor by many on the right, cruised to victory with 74 percent of the vote. Of course it should be noted that he was, you know, right about the whole weapons-of-mass-destruction thing, but still, he could have been dead wrong, run naked through downtown Seattle shooting random strangers, and eaten a baby koala— live on television—and he still would have received at least 62 percent. Seattle likes liberals.

It should also be noted that the Communist Party historically has always been strong in Seattle and I've heard we have one of the lowest rates of churches per capita among major cities in the nation. So if one were to claim that Seattle is a bunch of godless liberal commies, well, we would have to pretty much fess up to that.

This is the world I was raised in and where I've lived most of my life. Seattleites are aware that there are Republican voters that exist in the world, but those voters are sort of like those stars that astronomers can only posit the existence of, they cannot be picked up on any traditional viewing device. And yet . . . Sometimes, while reading *The Nation* and sipping on a latte, trying not to spill any on my Gore-Tex pullover, I would think about what *liberal* meant. I knew liberals were against the war in Iraq and against racism and homophobia and against Bush's tax cuts and against the power of major corporations, but what were liberals, you know, for?

I was also aware of the axiom that if you're a conservative when you're twenty you have no heart, and if you're a liberal when you're forty you have no brain. I couldn't help but wonder at age thirty-six if my liberal lifestyle was getting in the way of my natural evolution.

My life was not a conservative vacuum, however. My wife Jill's brother-in-law DJ is about as conservative as one can humanly get. The son of Dick Cheney's former chief of staff, he was once one of Ralph Reed's top men at the Christian Coalition and went on to be a Bush appointee in the Federal Highway Administration. DJ has his beliefs, he's sincere about them, and when we talk/duel/argue, those beliefs couldn't be more different from my own. Maddeningly, he invariably wins the debates we have. Too often, he has points while all I have are complaints. Of course, he has some rhetorical advantages since he earned a law degree from Georgetown while I earned a theater degree at an obscure liberal arts college, but the point remained: he won arguments.

Unlike the traditional liberal caricature of conservatives, DJ is a great guy. He does not secretly plot the conquest of the world with covert emissaries from Halliburton, he doesn't fly into a murderous rage at the mention of any member of the Clinton family, and rarely, if ever, does he roll around naked in mounds of gold coins stolen from third world families. He's a good husband, good father, and a patient golf partner.

Around the time of the 2004 elections, the program director at the public radio station where I work asked me to do more segments about national events on my weekly radio show. He thought it would be interesting to have a conservative and a liberal on together to hash out a particular question from week to week. "Is Iraq another Vietnam?" for instance, or "Should Rumsfeld be fired?" I was skeptical. "But won't that be an awful lot like those stupid shows where everyone yells and acts like jackasses?" Not if I didn't yell or act like a jackass, he told me.

So I tried it out and soon started having lengthy interviews with guys like Rich Lowry and Pat Buchanan. Shockingly, they turned out to be smart, friendly, helpful people who were articulate in their beliefs and formed their arguments coherently. I didn't always agree with them, but often they made a lot more sense than whatever was being argued by whichever liberal I was able to grab. Maybe it was the quiet confidence that comes from knowing their side was in power, maybe they were more personable because they were sitting in their luxuriously appointed offices, with overstuffed leather chairs, paid for by the Vast Right-Wing Conspiracy. A pipe full of good tobacco, smoked while not being encumbered by oppressive antismoking laws and hypersensitive liberals, and a snifter of brandy would make anyone a nice guy. Meanwhile, liberals sitting in stiff metal chairs in their makeshift storefront offices, constantly being detained and severely beaten by Patriot Act–enabled government enforcers, could hardly be expected to compete in the friendliness department.

But maybe, I thought, all those righties are confident because they're actually right.

As I reassessed my view on the conservative universe, I remembered Morgan Spurlock's movie *Supersize Me.* If all that McDonald's food was able to so radically transform Spurlock's body, what would a massive concentrated amount of conservatism do to someone's brain? Is liberalism like a liver or a kidney and will it just shut down after a while? Or could it be possible to switch? What if I could go over to the other side and, instead of merely appreciating and understanding what the conservatives have to say, really believe it and become one of them? Sure, some people drift from right to left or left to right over the course of years or after a series of randomly occurring cataclysmic events, but is it possible to change your own mind? Could I pull off artificial conversion?

What would happen if I invaded my own brain with troops in the form of conservative opinion, conservative experiences, conservative art and culture, and all the trappings of conservative life so familiar to red-state America and so foreign to me? Would those troops be greeted as liberators? Or attacked by insurgent brain cells? The more time I spent thinking about these possibilities and watching the postelection moping of every liberal I knew (parents at my son's preschool wept, actually physically wept, for weeks afterward), the more attractive it became to really try it.

With every Noam Chomsky lecture I heard on public radio, every "Lick Bush" bumper sticker I saw on an old Volvo, I wondered if I was like Neo in the movie *The Matrix,* trapped inside an illusory liberal universe waiting for Laurence Fishburne/George W. Bush to set me free. Would I get to learn kung fu fighting skills if I broke out of the Matrix?[1]

After some soul-searching and a somewhat awkward conversation with my deeply liberal wife, I requested, and was granted, a month's leave from the station. In that time I would change my wardrobe, travel the country on some carefully planned trips, and ingest all the conservative dogma I could as part of an effort to conservatize myself.

Yes, I risked my friends and family disowning me. But I would also have proven that people, even in this polarized America, really could change their minds if they heard something thoughtful that they had never considered. That people could be persuaded. That ideas still matter. The other thing that might happen is that Jill would likely divorce me and never let me see the kids again. That would be a drag, but I was sure that the military-industrial complex would gladly provide me with a new wife and shiny happy new children, possibly android in nature.

In planning the project, I needed some guidelines. While sitting at a Starbucks in Seattle with my Apple laptop computer, I jotted down some parameters to try to get to a place hitherto unknown by people in Seattle working on their Apple laptops in Starbucks.

### RULES OF THE
### SELF-CONSERVATIZING EXPERIMENT

1. No lies, no fake names, no deception. No stating an opinion that isn't really my own. I'm free to be cryptic about my opinions and turn questions around when asked them, but it has to be me going through this.

2. Activities are to be based on a purely unscientific but highly personal idea of what American conservatism means. While I will surely meet academic conservatives who have no use for country music and working-class folks who don't read *The National Review,* both of those things represent conservatism to me and so will be part of the research.

3. Throughout, I will be sleeping with a hot liberal woman, but as I've been married to her for ten years, that's grandfathered in.

4. On the issue of the president of the United States, I've always had a hard time putting the name George W. Bush after the word *president.* The 2000 election was highly controversial and 2004 seemed a little shaky too, in Ohio

especially. But during the experiment, he will be President George W. Bush. Full title. Every single time.

5. All news and information will be gathered by conservative outlets. No daily newspaper or radio from any source that has ever been accused of liberal bias (which is most of the news media I currently rely on). When something happens in the world, I will find out about it through the filter of conservatism. If Bush is caught selling heroin on the White House lawn, I want to hear how it was actually the fault of the degenerate liberal culture propagated by the Democrats in Congress who somehow forced Bush, against his will, into dealing smack and who have probably done things that were a lot worse. And then someone would bring up Monica Lewinsky. Nothing goes into my head without conservative context.

## Approved sources of news and information

**Fox News Channel**

**Conservative Talk Radio:** There are many programs to choose from. In Seattle, which would gladly elect Che Guevara as mayor if given the opportunity, there are two twenty-four-hour conservative talk stations. Access to these programs will not be a problem.

**WSJ Opinion Journal:** The actual *Wall Street Journal* provides plenty of ably reasoned and well-articulated articles about a variety of topics, and is assiduously skeptical of any issue. Meanwhile, the opinion page is full-on right wing. It's a place where Republican politicians are portrayed in much the same way that their dogs see them: flawless, noble, and beloved.

**The National Review:** Started in 1955 by conservative icon William F. Buckley Jr., the current version is a hangout of snarky modern conservatives. If the Kevin Bacon character from *Animal House* ran a magazine, this would be it.

***The Weekly Standard:*** Another conservative magazine, this one run by William Kristol, son of Irving Kristol, considered to be the founder of modern neoconservatism.

***Free Republic:*** Hard-core right-winger message board where ordinary people, unencumbered by editors or coherence, are free to blame Bill Clinton for every single negative thing that ever happened to anyone ever.

***NewsMax:*** News filtered first through conservatism and then refiltered through another process that packs every story with breathless hysteria. Picture Bob Dole after he's consumed eight Mountain Dews. It's like that.

***The Washington Times:*** All the goings on in the nation's capital, as told by conservative employees of the Reverend Sun Myung Moon, who says he's the Messiah. Religious conservatives, who have strong opinions about the whole Messiah thing, and think it was a different guy, are fond of this newspaper. I think they do this just to make my head hurt.

6. No talking politics with liberal friends. If the subject comes up, I must literally put my fingers in my ears and say la-la-la.

7. Music must be by artists known to be conservative, Republican, or sympathetic to those causes. Artists who have performed at either of President George W. Bush's two inaugurals are acceptable.

8. Movies will be gleaned from a list provided by devotees of FreeRepublic.com. They must either enforce conservative values or implicitly or explicitly endorse conservative beliefs or Republican policies.

9. Another rule and a big one: no arguing, only listening. It's easy when you hear things that you might not agree with to

dismiss them out of hand or find flaws in the argument or compose pithy withering retorts that you're sure will put the offending talker in his place. It's harder to just shut up and listen.

10. Drink Coors beer. The Coors family is famously blamed by liberals for everything from union busting to putting prospective employees through polygraph tests in the 1970s to determine if they were gay. Pete Coors, current family scion, ran for Senate as a Republican but lost when conservatives questioned the company's sponsorship of a gay-themed festival in Canada. Play with fire and you'll get burned. Play with gay fire and you'll get gay burned.

11. Steak whenever possible. Also beef jerky.

12. This all must take place within the space of thirty days.

This little idea had become a real experiment. I began referring to it as The Experiment, the capital letters emphasizing its importance. Summer came and it was time to get started.

After the haircut and follicular realignment courtesy of Julie the barber, The Experiment was only days away. My lefty wife had cheerfully told friends that if The Experiment succeeded she already had the divorce papers drawn up. It was her little joke. I hoped. Then there was the matter of my son. Just as I was preparing for a potential move to the right, my beautiful four-year-old boy, Charlie, was beginning his move from preschool apolitical bliss to the tree-hugging left. We had received a fund-raising letter in the mail from the Sierra Club, and because it had animals on it and Charlie loves animals, he wanted to know what the letter was all about. Jill told him that President George W. Bush and his friends want to drill for oil in Alaska and these people who sent us the letter want to stop them because they're afraid it will hurt the animals. "So," Charlie asked, "why did they send us a letter?"

"Because they want us to give them money so they can use it to try to stop Bush from drilling up there," Jill explained.

Charlie went right for his piggy bank, emptied out the eleven dollars he had to his name, and said "Here. I want to send this to them. I want to save the animals." And with that solemn pledge, my son became a member of the Sierra Club. They sent him a tote bag and everything. So as Charlie headed down the road of lefty activism, my hair, if we are to believe the traditional left = liberal, right = conservative analogy, was ready to join up with President George W. Bush, Jerry Falwell, and the *Wall Street Journal* editorial page. From the scalp up, I was already in. Now it was time to get dressed.

---

1. Political kung fu, I mean.

# 2

# The **Persuasive Powers of Shopping**

## and Allowing

# Lee Greenwood

## into **My Life**

☞

**In which the author encounters a confrontational hobo and prepares for potential self-alteration by acquiring appropriate wardrobe and accoutrements.**

Jill and I sat on a luxuriously padded bench in the posh men's suits department at the Nordstrom department store in downtown Seattle. Besides being my hometown, Seattle is the birthplace of Nordstrom, which existed originally to help outfit people who were heading off on great adventures to Alaska to search for gold. Today, it helps people dress for formal offices, formal events, and anywhere that calls for looking more important than people who shop at Target and Wal-Mart. The men's suits department is the store at both its hoitiest and toitiest. I was completely out of my element. A handsome suit is not a requirement at the radio station where I work, though in recognition of my advancing age, I make an effort (and sadly it really does require effort) to at the very least avoid T-shirts with band names or sports teams printed on them. While I used to have to throw on a tie during the occasional temp job a long time ago, the full-on grown-up suit was never required. When I later moved into an editing job in the dot-com industry and then public radio, well, you're lucky if guys wear pants in places like that.

But I could not imagine conservatives, especially the powerful ones who run the world and prepare the United States for an invasion of Venezuela,[1] would wear ratty old Seattle Mariners T-shirts. No, if I was to walk among the righties, perhaps become one, I needed to earn the instant trust of the powerful righty opinion makers with whom I would soon be meeting. I wanted the kind of thing Reagan wore, a suit you could wear while addressing the Republican National Convention or at a clandestine meeting of the Halliburton board of directors. Sure, I wanted to blend in and not be detected, but that was only part of it. More important, I wanted to feel what it was like to be dressed like that. Back in my time as an actor,[2] I was told by a costume designer that I shouldn't merely put on the costume; I should really *wear* it. Let it affect the way I felt as the character, let it become part of who I was inside. I didn't want to just buy a suit; I wanted to *own* a suit.

Then my thoughts were interrupted by a drunken vagrant who came shambling in out of nowhere. "What's the meaning of life?!" he coughed, getting a few inches from my startled wife's face. "I don't know," she offered gamely. "Why don't you tell me?" His pungent presence was a challenge to my biases.

*That man, reeking of alcohol, smelly, and looking as if he hadn't bathed or changed clothes in a month is really a good person,* I would have figured earlier in my life, *he just needs someone to listen to him. Yeah, okay, maybe he's drunk, but that's due to a substance abuse issue that society refuses to recognize and in the post-Reagan world there is no recourse for him to get treatment to deal with this disease that he happened to be struck with. And okay, sure, he's a little belligerent, leaning in on my wife a little more than I might like, but that's his anger over being marginalized by an uncaring society spilling over, and hey, you know, who could blame him?*

That's how I would have felt back then. This time I didn't.

Barely had my wife offered him her polite switcheroo question when I launched my response.

"Get the hell out of here!" I told him, "Leave us alone!" My tone was harsh and threatening.

He staggered back a step or two, then spat back, "What are you gonna do about it?!"

"I'm telling you to get out of here," I continued, looking him in squarely in the eye, "and I'm calling security. Excuse me?" I said to a nearby salesman. "This guy is hassling people in your store!" As the well-dressed Nordstrom sales staff scrambled to their house phones, the drunk staggered away from us. I had won.

I went hard line and it felt great.

I still think the guy needs help, I still believe that we need to treat people's chemical dependency issues instead of locking them away for decades in the name of "getting tough." I didn't want to see him put in a penitentiary for years, but getting liquored up and harassing people ought to be enough, in a civilized society, to get you kicked out of a store.

Now, dear reader, you might be thinking that The Experiment was beginning early. But I must confess that it was actually a continuation of a subtle rightward trend that I had been noticing in myself for a while. On certain social issues I had become, if not a square, squarer than I was. I saw the "Fuck Starbucks" decal on the espresso machine at my local independent staffed-by-undiscovered-rock-stars coffee shop and thought, "Gosh, do they need to use such foul language?" I would see teenagers with their pants hanging down past their bottoms and really wonder why their parents weren't doing something about it (and yes, I had come to refer to that body part as the "bottom"). And squareness means conservativeness. Sure there are some hip and cool conservatives, but on the other hand, no there aren't.

After the hobo battle, it was time to actually shop for the suit. An older gentleman named José soon approached us and before long we were shown a handsome navy-blue number. I was sent to try it on and then parade before Jill, who was comfortably seated in the mirrored lounge area. We felt like an inverted production of *Pretty Woman,* with Jill as the wealthy businessperson and me in the role of the adorable whore. The suit looked great, professional and elegant but also neutral and totally forgettable. Like former White House chief of staff Andrew Card.

José, a man so possessed of style that he made a wide-collared purple shirt look formal, asked why we were buying a suit that day. We explained The Experiment and talked about how I needed to appear conservative. "Ah yes," he said, "I do this all the time. I appear conservative here at work but I am still a liberal in my mind." He told us stories of his time spent in labor unions in New York City in the 1960s when he supported the liberal Republican mayor John Lindsay, and shared his thoughts on how he urges people who come to America from his native Puerto Rico and other countries to learn English. "If they learn English, they can make something of themselves," he said, "but if all they know is Spanish then they're just going to be housekeepers and dishwashers for the rest of their lives." I know that it is often the tactic of salespeople to tell the customer what the customer wants to hear, but I couldn't really figure out how this complex map of ideology really fit in with that notion. In José you had someone who supported a liberal

Republican and who pushed the English language as a means of empowering immigrants.

We picked up a couple of dress shirts, one white and one blue, along with a belt. José, Jill, and I scoured the shop for the most conservative necktie and Jill came up with the prize: a moderately wide number with bold diagonal red, white, and blue stripes that ascended upward as they moved from left to right. Perfect in so many ways. Final Nordstrom bill: a thousand dollars. I gulped at the cost but figured I could pay it off easily when I landed a cushy job at the Heritage Foundation upon my successful conversion. And it looked great. It was not just a suit. It was The Suit. Superman had his, which enabled a host of magical powers; now I had mine.

I was not going to be able to wear The Suit every single day, however. It would have looked fine when I met with pundits or even at the gun range,[3] but not so much at, say, a country-music concert. We left Nordstrom and set out to do a little more shopping. "I'm thinking we should do polo shirts and golf shirts, maybe a few pair of Bermuda shorts so you have that Republican leisure look going on," mused Jill as we drove toward the sprawling mega-Goodwill on Dearborn Street. "Then we'll head to Wal-Mart for the working-class conservative look."

"Well, that'll be good, to have a few things to wear to go along with my regular clothes," I added.

"What do you mean your 'regular clothes'?" she asked incredulously.

"Um . . . my clothes that I own . . . and wear . . . ?"

"No. John. These are going to *be* your clothes. You can't wear your regular clothes during this whole thing. Everything you own is liberal." My wardrobe flashed before my eyes. Carhartt work clothes even though I'm a lifelong office guy, Teva sandals, more than a few black T-shirts, worn flannel shirts that I never got rid of even after grunge became passé. She was right. Lefty McLiberal had compiled my wardrobe and had done so in, like, 1994.

Besides her being a wife, a mother, an artist, and the Mean Lady Who Took My Clothes Away, it's important to note that Jill is also an occasional costume designer for theatrical productions. I've seen her create garish pro wrestling outfits, moose suits complete with oversize papier-mâché heads, and a cocktail dress built out of license plates. I asked her what she was going for in the costume design of "John Moe, Conservative." "What I want," she said, "is for our friends to see you out walking in the neighborhood in these clothes and wonder if you're okay. I want them to come back to me and say, 'I saw John the other day and he was wearing these . . . *things*. Is everything all right with you guys?' " I got the concept even if I was a little alarmed by how lustily she regarded the appearance of marital strife.

Located at the epicenter of liberalism in the middle of Seattle, Goodwill was a gold mine of conservative gear that had been acquired (presumably as gifts from conservative relatives) and then rapidly cast off by the Chomsky devotee populace. I dove in and picked four polo shirts: a pale yellow one, a houndstooth one with the name of a telecommunications company, a blue-and-white-striped yellow number, and finally a sky-blue polo shirt with the logo of the Benny Hinn ministries on it.[4]

Meanwhile Jill was making good progress elsewhere in the store. She had retrieved a pair of pastel-yellow shorts that might as well have been yanked directly off the stiffening body of an old man living in a third-rate retirement village. She had gotten to work on the blue-collar ensemble as well, presenting a slightly faded Dwight Yoakam tour T-shirt and one that warned anyone who might be considering messing with Texas that that might not be a good idea.

A pretty good haul at Goodwill: six shirts, one pair of

horrible yellow shorts, total cost of maybe fifteen bucks. But while "Golf-Playing Conservative" was now nicely outfitted, "Conservative Groundskeeper at the Golf Course" needed new threads as well. We went one notch up the scale, to Wal-Mart.

Although it is the biggest corporation in the world, there are no Wal-Marts within the Seattle city limits. Not only would it be hard to procure the obscene amounts of land required for the store and parking lot, but the existence of such a store would offend to the core the liberal, anticorporate population of the city that still has no problem buying coffee at Starbucks on their way to their jobs at Microsoft and later shopping at one of several Target stores in the city. If a Wal-Mart were built in my town, it would be burned to the ground within moments. It would then instantly heal itself like in *Terminator 2.* Jill and I went to the suburbs.

Perhaps because it was summer and the Fourth of July was right around the corner, or perhaps this was just the way Wal-Mart was, but there was America stuff everywhere. We went for the T-shirts. The first was red with blue lettering spelling out 100% RED-BLOODED AMERICAN in all-caps just in case its message was too subtle. We also picked up a blue shirt with AMERICA emblazoned across in slightly raised puffy letters. Within the letters there was a picture of an eagle flying, in case you didn't get the idea of America from the word *America.* We were unable to find a baseball cap that succinctly carried the same brand of patriotic message as the shirts, but we did find the next best thing: a shirt that celebrated hats. On a blue background, it featured three ball caps with the slogans THESE COLORS DON'T RUN, U.S., and DON'T MESS WITH THE U.S. All of these America shirts were made in Nicaragua and cost six bucks each. We christened them Blood Shirt, Puffy America Shirt, and Hat Shirt.

While I failed to find an America hat, I did find one cap

that was appropriately red state and for only a few dollars. Red, pre-bent visor like the kids wear, with a billowing checkered flag and an enormous NASCAR logo on the front. It wasn't in support of a specific driver, just in praise of the organization itself. Rooting on a corporate entity. Also in the store were Rustler jeans, a knockoff of Wrangler jeans but cheaper, uglier, and more poorly made.[5] I was a bit conflicted by the Rustlers, ideologically speaking. On the one hand, they were a tacit endorsement of shipping jobs from an American company overseas, thus destabilizing the American economy, but also an example of free-trade capitalism at work. Then again, would providing jobs in Mexico mean that there would be less illegal immigration from Mexico (a long-standing conservative issue)? Or would low-paying jobs in Mexico frustrate workers there, thus spurring more illegal immigration than ever before? Tricky jeans. I agonized. In the end, I bought them simply because a modern blue-collar conservative would probably disregard all of those issues and simply buy the jeans because they were cheap and would work okay for a while.

Total cost of the Wal-Mart trip: about thirty dollars for NASCAR hat, Rustlers, Blood Shirt, Puffy America Shirt, and Hat Shirt.

Over the next couple of days, Jill went out and gathered the final components of the wardrobe. For the white-collar look: two pair of plain khaki shorts by Ralph Lauren, a handful of Polo brand polo shirts with the little pony on it and everything, slick black dress shoes to go with the suit, and some brown Top-Sider shoes that would look perfect at a golf tournament, fraternity rush party, suburban barbecue, or any number of other places where I traditionally feel hopelessly out of place and uncomfortable.

The ensemble was nearly complete, but the blue-collar look still had no footwear. On a subsequent afternoon doing errands in Seattle's Greenwood neighborhood, Jill spotted a thrift store and wanted to run in for a second. I waited in the car with the kids and she emerged a few minutes later with a bag and a great big smile on her face. In the bag was a pair of cowboy boots, big brown ones that looked like they had barely been worn. Also about a half size too small, as I learned while furtively trying to ram the size 11½ bastards on to my size 12 feet in the thrift-store parking lot. "Oh, we can get those stretched," said Jill, "they'll be fine." The next day we did get them stretched and they went from being unbearably painful to merely significantly painful. The wardrobe was complete. The preparation was not.

I've never specifically selected music based on the political stripe of the artist, at least not consciously, but it's worked out that when there have been political affiliations, I haven't exactly been snatching up CDs by the Pat Robertson Orchestra.[6] A lot of my favorite artists happened to have spoken out against the war in Iraq and expressed misgivings about George W. Bush, at least in the form of a passing comment when asked or tossing a song onto a compilation album. Prior to the 2004 election, my friends at McSweeney's and Barsuk Records released *The Future Soundtrack of America,* a compilation intended to benefit various liberal causes, including my son's beloved Sierra Club. I see myself as being open-minded to all sorts of music, but it was remarkable how much this record, this avowedly liberal record, resembled an index of my own music collection. On it was Death Cab for Cutie, They Might Be Giants, Tom Waits, Nada Surf, David Byrne, R.E.M., Sleater-Kinney, and Old 97's. Throw in the Free Tibet–supporting Beastie Boys, the Canadian Neil Young, and the anti–Tony Blair Robyn Hitchcock, and you would be well on your way to making me a mixed tape that I would treasure for a very long time.

Was my music making me liberal? It's easy to dismiss music

as mere entertainment, but that devalues its transcendent power, its spiritual place in our lives. People who listened to music like mine were more liberal people and people who listened to other types of music were more conservative. Where is the cause and where is the effect?

Since I couldn't imagine a life without music, I needed to reprogram the iPod. By the time I was through, I looked at the new playlist and made a sound that was equal parts gasp and whimper.

**Charlie Daniels Band**—*Essential Super Hits of Charlie Daniels Band.* I overlooked the wordy and redundant album title because there were songs I thought would be helpful, including "This Ain't No Rag, It's a Flag." In that one, Daniels explains that the flag is not, in fact, a rag, and that is why those of us in the United States opt against wearing said flag on our heads. Furthermore, he indicates, Americans have secured some guns and have decided to shoot those who have selected the rag-on-the-head mode of dress. I wonder which groups are targeted in Charlie Daniels's mission. Arabs? Sikhs? Dudes with mullets wearing bandannas at Charlie Daniels concerts?

**Clint Black**—*Greatest Hits II.* In the early part of Black's career, his records would battle those of Garth Brooks for the top of the country charts and Brooks would always come out on top. Clint Black is the Buffalo Bills of country music. He even married the mediocre TV actress Lisa Hartman. Their children will surely have successful lives where they don't quite ever reach the tops of their field.

But Black is a good businessman, and when war was looming in Iraq, he, like Halliburton, figured out a way to capitalize. He released "I Raq and Roll," an up-tempo country number that enthusiastically blurs the line between Saddam Hussein and the 9/11 terrorists. He performed at the first Bush inauguration.

**Craig Morgan**—*Craig Morgan.* This one was kind of a stretch but Morgan had one thing that set him apart: he's a veteran. Morgan was part of the American forces that went down to Panama to get rid of Manuel Noriega, a former CIA operative who had become inconvenient, and blasted heavy metal outside Noriega's residence. The invasion of Panama during the administration of the first President Bush, along with the Grenada invasion by President Reagan, were the sort of feel-good wars, completed rapidly and with great precision, that made lots of people think the modern war in Iraq would be easy.

Morgan wrote a song about his stint in Panama called "Paradise." In it, he talks about his preference for Tennessee over Panama and his fear of what he had to do but also how glad he was to have "won for Uncle Sam." The government told him to go, he went, and that was that. That's a kind of acceptance of authority often found in conservative circles and totally missing among liberals. I shudder to think what a liberal song about Panama would be.

**Daryl Worley**—*Have You Forgotten?* Worley was a midlevel country performer before the title song of this album made him big-time. It's a retort to those who opposed going to war in Afghanistan after the 9/11 attacks and one of the only songs to mention bin Laden (who, conveniently, rhymes with "forgotten") by name. It's kind of an easy song to go along with. It's also an easy song to answer. Have I forgotten about 9/11? Well, no. I haven't. Duh.

**Kid Rock**—*Devil Without a Cause.* Mr. Rock, who designates himself "Pimp of the Nation" in one song, found himself at the center of a controversy when he was invited to perform at George W. Bush's second inaugural. On the one hand, he was a favorite of the Bush daughters and in one song he posits, through imperfect rhyming, that it is time to rock when Old Glory drops. On the other hand, that song is titled "You Never Met a Motherfucker Quite Like Me," and some conservatives who are, you know, conservative, had a problem with someone like that being involved in their coronation. There are other tracks on Mr. Rock's record that obsessively discuss pornographic videotapes, sexual relationships with multiple partners, and his ability to procure sexual favors from women with large bottoms while others are left to work as pimps for Barbara Bush.[7] Conservative columnist Michelle Malkin led the ultimately unsuccessful charge to block Kid Rock's participation in the inaugural festivities. I bought the record. If all I had was country on the iPod I might go insane and that could hurt my chances of truly being converted.

Besides which, if it is a presidential inauguration, shouldn't the Pimp of the Nation be involved at least in a ceremonial capacity?

**Kid Rock**—*Cocky.* Hey, if I could get one Kid Rock record, why not two? And I was going to need something to counterbalance . . .

**Lee Greenwood**—*American Patriot.* Like the Baha Men with "Who Let the Dogs Out?," Greenwood, a former Reno blackjack dealer and lounge singer, catapulted to fame and fortune on the strength of a single song. In his case it was "God Bless the USA," a midtempo number filled with platitudes about America being good, free, and worthy of blessing. Unlike the Baha Men, Greenwood carved an empire out of that song. While he doesn't land on the charts anymore, he's always being called upon to sing it at ball games, ceremonial events, and at the Lee Greenwood Theater in Seiverville, Tennessee. Even for someone living in my liberal Seattle cocoon, this song has become familiar. Maybe if I gave it a few more listens I might see what all the fuss is about. On *American Patriot,* Greenwood reinforces his brand with songs like "America the Beautiful," "The Star Spangled Banner," and the popular chant "The Pledge of Allegiance," here put bizarrely to music.

**Michael W. Smith**—*Healing Rain.* Not only did Smith perform at Bush's inauguration ceremonies, he's also a Christian singer. I never listen to Christian music due to two big reasons: first, I'm more of an agnostic, and second, because it almost always sucks. But into the iPod it went.

**Toby Keith**—*Unleashed.* Liberals, I think it's safe to say, are often more reluctant to go to war than conservatives, at least in recent years. They'll argue about trade embargoes and diplomacy all day long, while conservatives are smearing their faces with camouflage paint and gripping a knife in their teeth. Enter Mr. Keith. In the aftermath of 9/11, he released "Courtesy of the Red, White, and Blue (The Angry American)," a song with a complicated title but a simple message: America is furiously angry over what happened and will put a boot in the ass of . . . someone. We're not really sure who, exactly.

Keith is not quite as simple as one might think. A little research reveals that he is a registered Democrat and supported Democratic Oklahoma representative Dan Boren in recent elections. But he also endorsed the reelection of George W. Bush, has claimed to be embarrassed by his own party, and generally describes himself as a conservative.

So my iPod was loaded, my outfits assembled, my leave of absence from the station approved, and my tickets booked.[8] I hoped I had everything I needed but it's hard to know what to pack when you're going into a great unknown. The first leg of this trip was to the political and philosophical hearts of conservatism: Washington, D.C., and New York. Enough dithering in Seattle and thinking about conservatives, it was time to meet up with some real ones.

---

1. Think about it: lots of oil, liberal leader friendly with Castro, matter of time.

2. A craft I studied in college and attempted with not much success to make my career before realizing the limits of my talents and ambition.

3. In an "off-duty FBI agent getting in some additional lethal training" kind of way.

4. Hinn's Ministry having made a fortune off of religious conservative donors, thus deftly combining the Christian right and unfettered capitalism.

5. A pair of jeans actually made in America would cost close to a hundred dollars, but the Rustlers, made in Mexico, were about eleven bucks.

6. No such orchestra exists. Yet.

7. It's unclear in the song whether Kid Rock means the famously hostile white-haired older Barbara or the occasionally drunken and dramatically more attractive younger Barbara.

8. The general tone among my friends when I was getting ready to embark upon The Experiment was something between bewilderment and panic. Never before have so many wished me such total failure in my efforts with such kindness.

# 3

# Why Must the Media Always Discriminate Against the Make-Believe?

☞

**In which our intrepid author embarks upon his fantastical journey and encounters a reporter who isn't really a reporter and finds the intersection of gay porn and hot tea.**

In the waning minutes before the first day of The Experiment, I was sitting in an airplane, on an all-night flight to D.C. Most of my fellow passengers were already dozing off. Me, I was cramming every word of an old issue of *Harper's* and the *New York Times Magazine* into my head while checking my watch. A kind of intellectual last meal before this all got started. Midnight passed on Alaska Airlines flight 26, ushering in day one. Richard Nixon's ghost didn't appear on the wing of the aircraft; Ronald Reagan's face was not mysteriously superimposed on that of the flight attendant; I felt no sudden affinity for Herbert Hoover. Being lousy at sleeping on flights, I fished Dinesh D'Souza's *Letters to a Young Conservative* out of my bag. If I were going to be any kind of conservative, after all, it would be a young one.

D'Souza had been one of the more junior members of the Reagan administration, where he served as a domestic policy adviser. Prior to that, he had gained attention as one of the founders of *The Dartmouth Review,* an alternative/conservative student newspaper (think *Village Voice,* then shrink it and switch all the points of view) at Dartmouth College. In its early-eighties heyday, *The Review* had a reputation for, depending on who you asked, exposing liberal bias at the Ivy League campus or cruelly harassing gay people. *Sodomites* was the term they used. Staffers once snuck into a meeting of gay students and subsequently published a transcript of what was said as well as the names of those in attendance, including some who were still closeted. Oh, and when some Dartmouth students erected a shantytown as a protest against apartheid, folks from *The Review* knocked it down with sledgehammers. From this paper was spawned D'Souza, now a respected author, Laura Ingraham, now a popular talk-show host, and presumably countless others who pursue their sledgehammering and gay bashing in more private ways or possibly in recreational leagues organized within their gated communities.

The book is structured as a series of letters to a fictional young man named Chris, who is trying to deal with the discrimination he faces as a conservative student and who is very interested in corresponding at length with Dinesh D'Souza. In other words, Chris is a virgin. To read D'Souza tell of his college days, they were just college kids rebelling against the system by having a few beers, arguing back with professors, and attacking minorities. Among my people, the things D'Souza and his colleagues did would be called "racist" or "homophobic." But in his world, joking that "Jesse Jackson may pull out of the presidential race this season. It has been revealed that his grandmother has been posing nude in *National Geographic*" is hilarious.

It was one in the morning on the flight from Seattle to Washington and I was already having my first crisis. I mean, college is college and we all did things that weren't on the up-and-up. I cheated on my girlfriend with my ex-girlfriend then redumped the ex to go back to the newer girlfriend, for instance, but I would never write about that in a book. Whoops. I read D'Souza for another hour or so. The book was mostly composed of familiar arguments about the size of the federal government (too big), affirmative action (not a good idea, hurts blacks more than helping), and America's founding fathers (almost homoerotically awesome). D'Souza was succinct but dull. And if D'Souza was a boring writer, I shuddered to think how boring this kid Chris must be to have become involved in such a correspondence. Weren't there keggers to attend? But then I remembered that Chris was fictitious, so, no, there was no one more boring than Dinesh D'Souza. I lapsed into a theta-zone half-sleep Dinesh D'Souza reverie.

The plane touched down at Dulles. I had been to D.C. a couple of times visiting family and seeing the sights, but this was going to be work. Hard work. Like coal mining. But mental. I walked by a Starbucks, and since I had a huge day ahead of me and was not going to get anything close to a nap, I craved a caffeine fix. It's not cool to admit to enjoying Starbucks coffee if you're an urban liberal. Starbucks is a corporation, after all, and therefore, goes the liberal reasoning, the coffee can't possibly be any good. I've always loved it and desperately wanted that murky goodness to shoot into my waiting bloodstream.

Still, I could not drink Starbucks coffee because even though it's a corporation, it's a liberal corporation. A hundred percent of the company's political donations go to Democratic candidates. Whether this is to better satisfy its urban liberal constituency or a sign of the political inclination of corporate leadership is up for

debate, but for my purposes, dickering over motive was imma-
terial. For the next month, no Starbucks. This was a fine rule
to make for myself before The Experiment started, but Dulles
Airport has a few Starbucks kiosks, and on little sleep, they were
hard to walk past. All I could do was stare forlornly as if I were
running into a former lover with whom I had had a deep, passionate, and close relationship. And that lover was a mermaid.
Who poured me coffee.

After stopping in at my in-laws to drop off some of my stuff
and grab a quick shower, I put on The Suit, stopping at the
mirror to affect a cold and frowny facial expression that com-
municated either "I have a serious and realistic approach to
the problems of the world" or "I'm constipated," then headed
up to the city. I had a couple of hours before my first appointment, so I walked over to Lafayette Park, across the street from
the White House, and sat on a bench. I gazed at the White
House and I tried to see it not as the seat of All Power in the
Western World but as a residence. Not the White House, a
white house. A house where some guy lives, who happens to
have a job, and that job happens to be president of the United
States. In that context, it struck me that the guy inside was trying to do what he thought was the right thing. This visualization was kind of working, so I took it a step further. It's easy
to question a lot of the decisions that the guy in the house had
made, but by all indications, he really believes those decisions
are the right ones to have made. And hey, sure, corporations
happen to benefit from the war, and yeah, they're more powerful than one might like, but if there were no corporations,
there would be no business in America and if that were the
case there wouldn't be much of anything and we would live
in a feudal agrarian economy and I'd be a serf, which would
suck cause I can't grow anything and animals generally hate
me. So here's this guy in a house that people take pictures of

themselves in front of and he's doing a job. Look, there's a light on in that room. Look, there's his lawn. I got a lawn. Looks like he doesn't have the same dandelion problem I have. Isn't that interesting? I've got more of a dandelion problem, and I have a wooden fence instead of a thick iron one, and he starts wars way more often than me. Some dude in a house. Just this dude in a house. Finding myself gazing slack-jawed at the White House, I just kind of zoned out on that one for a while.

After this, the weirdest meditation session I've ever had, I looked at my watch and realized it was time for my meeting with Jeff Gannon.

By the time we met, 2005 had already been quite a year for Gannon. He was working as a sort of reporter in the White House press corps, going about his job in the best way he knew how, when, in January, he started receiving some scrutiny. Talon News, for whom Gannon was the only reporter, was owned and operated by GOPUSA, an activist group that says on its Web site that its mission is "to spread the conservative message throughout America." Gannon was known to ask questions of Press Secretary Scott McLellan and occasionally President George W. Bush that didn't exactly challenge the administration's talking points, most notably this one, asked of the president, "Senate Democratic leaders have painted a very bleak picture of the U.S. economy. [Senate Minority Leader] Harry Reid [D-NV] was talking about soup lines. And [Senator] Hillary Clinton [D-NY] was talking about the economy being on the verge of collapse. Yet in the same breath they say that Social Security is rock solid and there's no crisis there. How are you going to work—you've said you are going to reach out to these people—how are you going to work with people who seem to have divorced themselves from reality?" It is safe to say that Gannon's approach is not held up as an ideal by most journalism schools.

Rush Limbaugh would go on to claim that Gannon's question was inspired by Limbaugh's own program since Limbaugh had discussed Reid's use of the term *soup lines* even though Reid had never said that. But, said Limbaugh, "that is my characterization of their portrayal of America." So Limbaugh's untruth on his talk show was morphed into a presumption by a fake reporter from a news organization that served as a wing of a group called "GOPUSA" and then asked of President of the United States George W. Bush.

Tradition among reporters tends to favor tough questioning over sycophancy, and after a while Gannon's colleagues began to wonder precisely what the hell he was doing there. The political blogs picked up on that question, and as one can easily do online, they started digging. The series of exposés that followed revealed not only the thinness of Gannon's journalistic experience compared to that of most on the White House beat, but another kind of experience not often found among Washington reporters: gay prostitution. He had registered Internet domain names such as hotmilitarystud.com[1] and there were online profiles with his photo that appeared to offer his sexual services.[2] By the time I stopped looking into Gannon's background, I was still unclear as to whether he had been an actual man whore, a model for online porn, or some kind of smutty webmaster, but I still reached the conclusion of "ick." Also revealed: his name is not Jeff Gannon. It's James Guckert. Dude was using a fake name. Despite all this, he had received White House press credentials, and was permitted to attend scores of press briefings and press conferences over the course of two years.

As the heat turned up and the "who is this guy" factor built to a roar, "Gannon" bailed. He resigned from Talon News, which GOPUSA eventually stopped operating under that name (a visit to the Talon News Web site in October of 2005 revealed a Web page saying that "Talon News will be offline while we redesign the web site, perform a top-to-bottom review of staff and volunteer contributors, and address future operational procedures"). Later, "Gannon's" (fake) name would be connected to the Valerie Plame scandal.

I wanted to meet with "Gannon" for several reasons, none of which involved hot sex. I had already arranged meetings with several notable conservative pundits who'd gotten to the top of their profession through hard work, intelligence, and connections. "Gannon" had none of that but still made it to a point in life where he could ask questions directly of the president. When he did have that chance, he chose the most ass-kissy questions he could. If conservatism is a drug, "Gannon" had some access to a really good stash. I wanted to get me some.

How would I track down this elusive shadowy figure, scorned by liberals and disowned by conservatives, and get him to open up? Here's what I did: Googled "Jeff Gannon," found his blog, wrote in using the listed e-mail address explaining The Experiment and asking to meet. A couple of hours later he wrote back and said sure. We arranged to have tea.

He was a really nice guy. Pleasant, cheerful, well groomed. The liberal within me would have asked him why he was serving as a puppet for a corrupt administration. The newshound within me would have pressed him to try to establish a concrete, verified time line of his rather curious past. But for the purposes of The Experiment, I was there to learn about what made him a conservative. I expected "Jeff" would be nervous and evasive about his story. *Au contraire.* Since everyone else had been labeling him, I began by asking how he would describe himself.

"A colorful journalist with a colorful past," he said, "and one has nothing to do with the other. And they've used my past, and some things are true, some things aren't true, to undermine my credibility. The essence of the story is that they tried to discredit my work, and when they couldn't discredit my work they had to discredit me personally, which has taken the politics of personal destruction to a new level."

In fact they had discredited both his work *and* him personally. Many Talon News "stories" were revealed to be mostly cut-and-paste jobs from White House press releases and AP wire reports. The White House press corps and the liberal media decided, according to "Gannon," that they didn't like having a conservative in their ranks. "They poked around at my work for a couple of weeks: 'Who was this guy? How did he get here? A guy like him shouldn't be there.' Well, what do you mean by that? A conservative couldn't be within the sacred confines of the White House briefing room, which is liberal territory?" Regardless of the reporters' political inclinations, the people in charge over there were certainly not lefties. I asked why, if the White House was made up of conservatives, he didn't remain in the good graces of the administration and ride out the criticism.

"I resigned on February nineteenth because there were threats being made against myself, my person, and my family. I can take threats against me. People were going to mug me, steal my wallet to find out my real name. They were going to create an accident with my car because they knew the neighborhood, so we would have to exchange e-mail information.[3] I had people stalking me. But when it gets to the level of threats against my family, then . . . it was so important for them to drive me out of the press corps, that's an easy choice for me to make. They harassed my seventy-two-year-old mother, who's a widow, lives alone in a rural town hundreds of miles away from me, and they contacted her. Now, when your mother calls you up and says, 'Who are these people, what's going on, are they going to hurt me?,' what do you do? I felt that resigning and making it known why I was resigning would stop that. And it did for the most part."

I wanted to stick with the conservatism but I was getting sucked into the soap opera. So what was the deal with the fake name?

"Back when I started writing, I knew what I wanted to do. Everybody on the right who has an opinion wants to be the next Ann Coulter or the next Rush Limbaugh or Sean Hannity. So I wanted to choose a name that had commercial appeal. My name is very ethnic. And of all the people who talked about me on television, not a single person has pronounced it right. It's a very hard German name. It should have an umlaut on it."

This was a line of reasoning I had not heard before and one that quickly made a lot of sense: he wanted to be a pundit. He wanted to host talk radio and have an eponymous show on Fox News Channel and a book with a picture of him smirking on the cover. In his reasoning, the name would get in the way of that, even though the gay porn thing would somehow not. And he had another reason for changing the name from Guckert to "Gannon": mockery. "*G-u-c-k,* how many times growing up do you think I heard that in all kinds of obscene permutations?"

"Try having the last name Moe," I told him.

"The White House was fully aware that I used a pseudonym. When I submitted my request for the pass, I give my legal name, Social Security number, and that's the ID I present at the gate," he said, lapsing into the present tense as if not quite letting go of it all. "And then I slip on my Jeff Gannon badge. It's entirely ethical and there was never any attempt to do anything more than just keep a commercial-sounding name."

I asked him if he had any experience in journalism before getting this gig. "Not professionally. But all these people who say I didn't have any journalism background, I had more journalism background than some of my colleagues. I actually started my high school newspaper; I was the sports editor of my college paper and wrote a significant number of articles for them for two years."

He was not joking about these being legitimate preparation for a White House reporter job. He was completely serious. I stared at him for a long moment and then took a long sip of tea.

Finally, I asked how he actually landed the job. "I had been writing opinion pieces, submitting them around the Internet."

"Like on message boards?" I asked.

"Yeah, stuff like that. I had one Web site called The Conservative Guy. And [Talon News] was looking to add writers. They wanted to create a national news service and it wasn't what I wanted to do but okay. I submitted writing samples and submitted sample articles. It's not that big a deal. With all due apologies to these Harvard and Columbia graduates who are still paying off their loans, it's not that big a deal."

These were interesting points, tinged with delusion and madness. I switched up my approach and asked about the parts of his story that did not involve the White House or his status as a "hot military stud." How, I asked, did you become a conservative?

"I grew up in a rural part of Pennsylvania. My family is Roosevelt, Kennedy, John Kennedy Democrats. They're union people. And that's good. Because at one time, there was a certain element of the Democratic Party that was seen as representing working men and women. National defense, patriotism, those kinds of things. It was only in the sixties, when the Vietnam War was winding down was really when the country started to split. And that's when you start to see the development of the people that control the media today. I agree with people like Rush Limbaugh who say that they're more the public relations wing of the Democratic Party. These people in the media stopped being objective reporters of facts. They were activists. They were pro–civil rights, antiwar, blame America first. They were anti-Nixon. They were marching in the streets then; today they're running the press rooms."

"Gannon" had not yet been born but Guckert worked as a teacher briefly and then drove a beer truck before becoming part of the trucking company's management structure, where he got disgruntled with the unions and then became a conservative. "Do I have time in the next month to get hired by a beer distributor?" I wondered to myself.

I asked him if he had any advice on my project.

"Live your life and think. Be thinking. Be a truly thinking person. Not an academic, but a truly thinking person, who looks at the world around them, will, over the course of time, develop conservative ideals if they don't start out that way. I know of very few people who were conservatives and decided to become liberals. It's always the other way around. The realities of life give people the basis for adopting conservative values." Apparently the problem with Howard Zinn, John Kenneth Galbraith, and Bill Clinton was that they had simply not thought things through.

"What are those values?" I asked.

"Strong national defense. There's something great about America and it needs to be protected at all costs. The notion that there's some sort of moral equivalence for our nation's enemies is wrong. The idea that there were no terrorists in Iraq, no terrorists in Iran, or Saudi Arabia or Syria or Lebanon, is ludicrous."

Okay. America is great. Military, besides being hot and studlike, is important. "So why don't we attack Saudi Arabia or Syria or Iran?" I asked.

"We're not done with this war on terror yet," he said.

"Do you think that we should go attack those other countries?"

"I think it's probably good that I don't make those decisions on behalf of our country."

And with that bit of agreement having been established, I drew our conversation to a close.

Still, I had arrived at an interesting lesson. If my ambition was to become a conservative pundit/talk-show-host/gadfly, I had learned how not to do it. Because I think that's why Guckert became "Gannon." He wanted to be famous and respected. He figured if he changed his name, dressed nice, and genuflected to the Republican president, all that other business might never come up. The problems were that in the age of Google, *everything* comes up *all the time* and that conservatism, social conservatism anyway, is built on the idea of being none too keen on homosexuality, let alone homosexual military-themed prostitution on the Internet. Having never engaged in such activities, at least not as far as I can remember, perhaps the door to the conservative punditocracy was indeed open to me should this whole thing work out.

I bade farewell to "Jeff Gannon" and watched him saunter off into the D.C. afternoon, impeccably dressed, relentlessly cheerful, and wholly convinced that his college-paper sports editing and message-board portfolio had made him perfectly suited for White House correspondent duty. I could see the appeal of being a Bob Novak or David Frum or even Michelle Malkin.[4] Whenever I see *Crossfire* or those other shouty cable-news debate shows, I'm struck by how serene the conservatives are. The liberals get apoplectic, the conservatives appear smugly satisfied.[5]

Walking through Washington, heading back to the Metro train that would take me back to my sister-in-law's house, I happened to stroll by the headquarters of National Public Radio on Massachusetts Avenue. For those of us who've spent much of our careers and given up hope of a decent salary in support of careers in public radio, the NPR building is kind of a shrine. So is the network. On the few occasions when I've had a commentary aired on *All Things Considered,* I get all goose-pimply, not just because I'm nationwide but simply because I get to hear Melissa Block say my name out loud. I wanted to go in. I wanted to meet Robert Siegel and Renée Montagne and Michelle Morris and Steve Inskeep and all those names I knew so well. I felt like a Mormon standing outside the temple in Salt Lake City. I didn't care if I just got to sit in the lobby by myself, I wanted to be . . . of . . . it. But some people say NPR is rife with liberal bias, so I couldn't go inside. "I bet they're all in there right now being probing and insightful," I sighed to myself. "Maybe they just now made a subtle point or offered a perspective that one might not have previously pondered."

As I stood there, I sang the *All Things Considered* theme to myself: "DA da da da da da da da, da da da da da da DA!" I turned away, put on the iPod, and fired up some hits by country legend and right-wing extremist Charlie Daniels.

---

1. Although he later claimed to have done so for a "client."

2. "8 inches, cut."

3. Armed with his e-mail address, "Gannon" figured, the car-crashing liberal stalkers would suddenly have tremendous power over him. It seemed a bit odd to me that he imagined this horrifying car-crash/e-mail swap scenario wherein people would try to get his e-mail, when I had found it online in about twelve seconds and e-mails of reporters are often printed in their stories. Besides, aren't you supposed to actually give out insurance information as opposed to personal contacts?

4. Novak politely declined my interview request, Frum and Malkin never responded at all.

5. Novak being the exception when he went absolutely bananas on James Carville during a CNN appearance and was subsequently fired. But among commentators not caught up in an administration-led campaign to reveal the identity of CIA agents as an act of political retribution, there is a noticeable sense of calm.

# 4

# Ent

t

S

**In which the author sets forth to**
**debauched metropolis, traveling by rail, and**
**is offered a furniture-inspired metaphoric frame**
**of reference for viewing the modern right.**

On the Amtrak train from Washington to New York, I had a few more hours to get to know the music I had programmed as part of the effort to reprogram myself. As I've mentioned, the right lacks the broad spectrum the left has when it comes to popular music. We have everyone from Joan Baez to Green Day to Springsteen to the Dixie Chicks to, well, pretty much every other artist in the world. The right has a smaller selection, a great deal of which blows. There's Lee Greenwood, who presents a very traditional, "as seen on the-Hallmark-Channel" sound with melodies that will easily stir the emotions of people whose emotions are easily stirred by things like Lee Greenwood melodies. Michael W. Smith is a Christian and likes to let everyone know about it, as Christians sometimes do. He sounds like the youth ministry coordinator who "just wants to rap at you kids for a while." Smith sings songs that are either about his wife, his girlfriend, or Jesus, though it's not always easy to tell which. Kid Rock has a song where he orders you to put his balls

. Conservative music really is a big tent. I mused
hile traveling the five states one goes through on the
from D.C. to New York (six states if you count D.C.
but conservatives have traditionally opposed the D.C.
hood movement and so I wasn't sure if I could count it or
t). I also had some reading to get done before the train pulled
into Penn Station. *Legacy: Paying the Price for the Clinton
Years* is a lengthy analysis of the Bill Clinton administration by
Rich Lowry, who has been the editor of *The National Review*
since 1997 and is also a frequent contributor to the Fox News
Channel. To understate it dramatically, Lowry thinks that Bill
Clinton was perhaps not the greatest president we've ever had.
Folks on the left often blame Bush for 9/11, using the whole "he
was president when it happened and received urgent warnings"
argument. Lowry goes further back. "[Clinton] knew about the
threat. Yet, he tolerated a terrorist sanctuary in Afghanistan,
tolerated Saudi and Pakistani support for terrorism, put the
FBI and the CIA in the position of always playing defense, and
made the protection of American lives from terrorist attacks
no better than a secondary priority of the administration. The
September 11 attacks finally gave Clinton the kind of legacy
he had yearned for: one that couldn't be ignored, one that was
great in its implications. His could no longer be considered an
inconsequential presidency."

Meanwhile, liberals tend to love Clinton in the same way
that conservatives love Ronald Reagan, each side lining up
behind their eye-twinkling two-term chief executive, overlook-
ing the hero's flaws and focusing on the warm loving feeling
the icons inspired. Likewise, conservatives attack Clinton with
much the same fervor as liberals attacked Reagan, especially
during that whole "what if he blows up the world" period in
the eighties.

But I've interviewed Rich on the radio and I like him. He's

smart, well-spoken, and unfailingly polite. Best of all, he's conservative without being stodgy. In an interview about Bush's nomination of John Bolton, the American ambassador to the U.N. who had a reputation for having the temper of Mike Tyson and the face of the dude on the Pringles can, I proposed that perhaps the nomination would be better served if Bush had nominated Michael Bolton instead. "Well, that would create a whole different set of problems," Lowry glibly retorted without missing a beat.

I found the approachability of his style to be potentially useful for The Experiment. No one has ever been won over by the shriekers on the right or the left. Such figures exist, in large numbers, solely to rally the troops, to give the choir a nice sermon, and for people on the other side, they repel rather than attract. I had asked Rich if he would be willing to try to convert me, to help the scales fall from my eyes so I could see the deep abiding wisdom of the conservative point of view. Sure, he said, he could give me half an hour. Not sure if that's all the time he had to spare or if that's all he thought it would take but regardless, I traveled to *The National Review*'s offices on Thirty-second and Lexington Avenue to find out. Once again I was wearing The Suit. Given that it was late June and I was in New York City, The Suit probably would not have been my first (or 185th) choice but I had to get the conservatives to accept me as one of their own. Like Jane Goodall, I had to earn their trust. If they had been baboons, I would have painted my bottom in vibrant colors.[1]

*The National Review* is housed in an unremarkable midtown office building where it takes up part of the fourth floor. On first impression, the space reminded me more of a fly-by-night telemarketing operation than the headquarters of a magazine widely feared and loathed on the left. The young receptionist, I'm guessing a recently minted college conservative who had demonstrated facility with sledgehammering and was here as his stepping-stone to righty fame and fortune, ushered me into the library. It didn't look so much like an austere and well-appointed reading room as what an austere and well-appointed reading room might look like if it were part of a set built for a college theater production.

The walls were cheap wood paneling, rec-room style, the shelves lined with various conservative tomes from down through the ages. Regnery, the publishing house that has done exceptionally well bringing the public books that span the ideological spectrum from "conservative" to "very very conservative," was well represented. There were also older titles, including, prominently, everything William F. Buckley Jr. has ever written. A number of the books appeared to be arranged and labeled in a sort of halfhearted attempt at a Dewey Decimal System; others were just sitting there. This was the source of some of the most important conservative opinion in America and therefore the most relevant ideas in the modern world? "Would you like the latest copy of the magazine?" the boy receptionist asked. "Hot off the presses!"

"Oh," I said, mustering an enthusiasm that I hoped would eventually become real, "great!" Before I had a chance to really start reading, Rich Lowry came into the room. No suit. A shirt, slacks, no tie. I felt like someone who had overdressed for a job interview. I asked him if he thought the goals of The Experiment were possible. Could I change my own mind?

"Sure, it's never too late. It has a lot to do with who your parents are, where you started, how you're raised," he said cheerfully. I flashed back to my antinuke-activist sister who finished college right as Carter was leaving office and my brother who wrote to the Selective Service that he refused to register for the draft because Ronald Reagan's foreign policy "sucks donkey balls." My parents were native Europeans and former theater folk. My DNA was not going to be much help in The Experiment. Uphill battle.

I asked Rich what got him started. He said it was the 1984 presidential election, when he developed a bit of a political man-crush on Ronald Reagan. "It wasn't really rational. He just seemed a really attractive figure with great strength. Patriotism, optimism. I worked my way a little backward, [asking] 'What does he believe? I want to believe what he believes.' " I guess I was hoping to hear something about tax cuts or foreign policy, but apparently the initial draw was Reagan's hunkiness, politically speaking.

1984 was the race against Walter Mondale when the entire country, Mondale especially, knew from the beginning that Mondale had no chance. The nomination was a gentlemanly reward for a party loyalist who was given an opportunity to hold the spotlight for a few bittersweet months in a doomed run against a popular incumbent. I like to think Mondale and Bob Dole have talked about this over cocktails. In 1984, Lowry was a high school sophomore; he graduated in 1986, the same year as me. Once he reached the University of Virginia, Rich Lowry started a conservative student newspaper, *The Virginia Advocate,* modeled after the antigay vigilante-mob/newspaper *The Dartmouth Review* that Dinesh D'Souza had been operating. Lowry had read about *The Dartmouth Review* in *The National Review.* And now he's here, running the latter. So you see, it all connects in tight little concentric ways.[2] It bothered me a bit that someone my age had risen to so much more power than I had, but it also gave me hope for The Experiment. Conservatism wasn't solely for jowly old establishment hacks or the wild-eyed President George W. Bush Youth Brigade that helped shut down the Florida recount. It was for people of my generation too.

At this point in The Experiment, I was still a little unclear on what conservatism, you know, *was*. I knew it was the opposite of liberalism, but since most of liberalism in recent years has been about opposing conservatism, it all got quite foggy. There was something about guns and invading Iraq, but the rest of it was an ideological chowder in my mind with pro-life marches, intelligent-design textbooks, golf, and big fences to keep out immigrants.

So Lowry explained three basic tenets:

- Libertarian free-market economics with small government and low taxes;
- Social conservatism, which seeks to preserve the fundamental mores and traditional values of society;
- Strong aggressive defense.[3]

"Those are the three legs of the stool," he said. "Occasionally you'll see someone wanting to take those out and we always want to keep them together because otherwise you don't have a majority coalition. John McCain doesn't want the religious conservatives in there. Well, that would be a huge problem. Pat Buchanan doesn't want the free marketplace in there. Well, that would destroy it."

As happens with most political conversations, we quickly turned to the subject of the war in Iraq and he said conservatives had supported the war for several reasons. "One is just the threat Saddam presented in that strategically crucial region and the belief, widely shared, turns out to be wrong, that he had weapons of mass destruction. Post 9/11 environment [there were] worries about connections he might have to Al Qaeda. There's a huge theological debate about whether he had connections and how active they were. There were some. And the third, and this for me has a slightly Wilsonian aspect to it, this is the cauldron of anti-American hatred in the Middle East. And Islam is not the problem. It's the intersection of Islam and this poisoned political culture in the Middle East that creates the problem. So what you have to do ultimately is clean that up. And Iraq is a central country there. The three key capitals historically have always been Cairo, Damascus, and Baghdad. If you have a Baghdad that's decently run, decent government, not perfect, but is kind of a model for the region, you can kind of change the whole geopolitical tilt of the Middle East. And if you succeed in that, you're going to be safer."

So, I asked, the administration thought that the whole "invade Iraq and make it stable" thing was going to just . . . work out?

"Don't let Iraq—things haven't gone well in Iraq but a lot of it has to do with the practical details of the implementation. I wouldn't let that turn you off to conservatism entirely. There's a myth that there was no planning. There was planning, it was the wrong kind of planning, and it was poorly coordinated within the government."

It struck me as illuminating about the situation in Iraq that even defenders of the war were compelled to declare the existence of planning itself.

Turning to domestic affairs, I said I felt like the American dream is just plain out of reach for people and they can't seem to crawl out of the cycle where they work hard, pay taxes, do everything they're supposed to do and still remain poor and trapped. They're hosed. Even I, by no means poor, have to work two jobs to pay my mortgage on my relatively small house and support my family. I'm lucky enough that my second job is free-lance assignments that I can do on my own time and not delivering pizzas or cleaning hotel rooms, but it still always works out to a lengthy workweek. Meanwhile, my parents bought a four-bedroom, three-bath split-level house with a huge yard in the Seattle suburbs in 1971 for $31,000. You're going to have a hard time finding any kind of three-bedroom, two-bath house without a yard in that community for less than $310,000 today. Meanwhile, wages have certainly not multiplied by ten.

"So you feel like you're getting hosed too?" he asked.

"No, I'm not getting hosed," I replied. "Maybe I'm getting, like, sprinkled. But it's hard for lots of people."

"We have this huge government," he said, "and, you know, one of the reasons everyone is feeling a little hosed or sprinkled is they're taking forty percent of your income, and it makes it harder. And some of that goes to help needy people or poor people and I would be perfectly willing to cut some sort of bargain where we increase programs for the poor if we stop doing all the stuff that's ineffective or goes to subsidize people who don't need subsidy. The farm program goes overwhelmingly to corporate operations. The Medicare prescription drug benefit is going to be hugely expensive. If that was going to be focused on people who can't afford their drugs, okay. But it's going to everyone, because [the elderly are] such an important constituency that everyone in Washington, Republican or Democrat, wanted to play to. Social Security, a lot of it goes to people who need it, some of it doesn't. So if we had a state that was focused on having safety nets for needy people, I think we'd have a much smaller government."

"Oh," I thought to myself, and paused even in my own brain, "that actually makes a fantastic amount of sense."

Rich continued. "I think most of the people getting left behind, it has to do with education. If you graduate from high school, if you get married, and you don't have children out of wedlock, the chances of you being in poverty are almost nil. But we have huge swaths of our cities where that does not happen and it's just the opposite that's happening. We need to find ways to combat that scourge. When Bush talks about promoting marriage, I don't think there's anything in liberalism as such that should react against that."

Lowry said he endorses the idea of government initiatives aimed at holding families together, especially when there is economic hardship putting a strain on marriages.

This struck me as an interesting combination of big intrusive meddlesome government and the protection of social mores. Stool leg #1 battling stool leg #2. And also, does anything sound more misguidedly liberal than "marriage counseling from the government"? But you pay for it, he said, by slashing pork from other parts of the budget. Trim up the odd farm bill here and there and you're on your way to a smaller government as well as helping poor families crawl out of poverty and participate in the American dream. An attractive idea and also, with its small government and traditional family notions, a conservative idea.

Part of President George W. Bush's plan for curing America's social ills has been the concept of faith-based initiatives, wherein church organizations would take a little bit of government money and help drug addicts and troubled youth and then throw in the Jesus (or Buddha or Allah or Krishna, but, who are we kidding, Jesus) for nothing. It's long bothered liberals because of how Jesus-y it all sounds since they get pretty fired up about the separation of church and state. I told Lowry that the counseling thing sounded interesting but I feared it would get pawned off on the churches, who would use a counseling style that is less clinical and more "believe what we say or burn in hell." He never even argued this one. "The faith-based stuff, we've been against," he said. "We think it's bad for the faith-based groups to get entangled with government."

"Oh," I said. "So maybe I don't have to learn to like that idea. That saves me some time."

"If you go around and talk to enough conservatives, you can pick off one liberal thing they agree with liberals on."

There were plenty of ideas kicking around my head by this point, but I wanted to get back to the special feelings Lowry had experienced with Reagan and whether he felt the same about any prospective 2008 presidential candidate. Was there someone to get on board with now so I could already be a dedicated follower by the time they got big and everyone loved them? Like getting into R.E.M. with the "Chronic Town" EP?

No, he said, no one really excites him for president at this point. He said Florida governor Jeb Bush is a bright guy, one of the smartest we have out there[4] but everyone says Jeb's not running including Jeb, so that's not going anywhere. John McCain doesn't do much for Lowry, and with his casting aside the religious right, McCain has jeopardized one of his legs and created a faulty stool for himself. But regardless of who gets the 2008 nomination, Rich Lowry feels confident he'll be supporting that person in favor of Hillary Clinton, whom he feels is guaranteed to be the Democratic choice.

Hillary. The not-quite-a-full-term junior senator from New York. The former first lady. The one that everyone on the right seems to hate so very, very much. Call her what you will, Hillary Rodham, Hillary Clinton, Hillary Rodham Clinton, Hillary Rodham Clinton Cougar Mellencamp, or simply Hillary, she will come up in a conversation with a conservative. You could make drinking games around it. Meanwhile, no other Democrat will necessarily be discussed. You could go for months without thinking about John Edwards or Kerry; Joe Biden is highly influential but rarely considered in conversation; Evan Bayh is a probable presidential candidate but unknown to anyone outside a plurality of Indiana voters.

Lowry said that for many people Hillary seems kind of power hungry and people don't like that, while also allowing that "when it's someone on the other side, they're power hungry; when they're on your side, they're ambitious. And she's just not a warm personality."

"Not like Dick Cheney 'cause he's a really wonderful guy," I said.[5]

"He has a sly warmth," Rich protested meekly.

I had been told that I was to have no more than thirty minutes with Rich Lowry, but I had been there for over an hour. Perhaps he wasn't as busy as his assistant had told me; perhaps conservative pundits earn a small commission if they really do convert someone.[6] Or maybe for Rich Lowry it was simply a pleasant conversation. It was pleasant for me. *The National Review,* and particularly its online group blog, The Corner, is often criticized as a viper pit of nasty neocons that serves as an echo chamber for, and takes marching orders from, the larger conservative power structure. I still can't guarantee that that's not the case, but I had spent a big chunk of the afternoon exchanging ideas, thoughts, and opinions with the guy who runs the magazine and everything he said made sense. Plus, you would think that if the larger conservative power structure was really pulling the strings, they would have sprung for a part-time librarian to fix up the place.

Walking out onto Lexington Avenue, I strapped on the iPod as I headed back to my hotel on Eighth Avenue and Forty-third Street. Times Square has been cleaned up in recent years for the family-friendly purposes of the tourist industry. It's kind of the hip thing to complain about the scrubbing clean of the once unscrubbed chunk of midtown, to bemoan the Giuliani-guided development of chain stores where once stood porn theaters. But I'd lived in New Jersey for a short time before this scrubbing took place, and whenever I found myself in Times Square during that time, I would feel both excitement about being at the most famous intersection in the world and enormous fear that I was about to be stabbed in the abdomen. The nascent conservative in me felt that if some liberal degenerate really wanted scuzzy Times Square porn, it was still readily available over on Eighth Avenue.

I switched on the iPod's shuffle option and along came Michael W. Smith's "Healing Rain." It's about Jesus, whom Michael W. Smith thinks is terrific. Contemporary Christian music is about the only place you can hear someone sing about leprosy with a pop-synth beat behind them.

Back at the hotel, I checked to see what was on *National Review*'s The Corner. Nothing from Rich about me and how awesome I was, but there was this entry from *NR* regular John Derbyshire:

> **Paul Johnson, in his History of Christianity, actually makes a case for the Spanish Inquisition, and I have never been able to think too harshly of them since. (In brief, and from memory: Medieval society was very disorderly, and a heretic wasn't just someone who had a polite disagreement with you, but more likely a rogue sub-intellectual type— think Ward Churchill—with a head full of half-digested theology and a rabble-rousing oratorical style. In a society with no police force or standing army, and a large floating population of unemployed peasants & craftsmen, a rabble once roused could level an entire city and cause untold misery. As a force for order against chaos, the Inquisition was generally popular . . . )**
>
> **Posted at 04:55 PM**

Okay, so conservatives wanted to help the marriages of poor families in the inner city but they're also fans of the Spanish Inquisition. I had a lot to learn.

1. Jane Goodall never painted her bottom in vibrant colors, as far as I know, but you get my point.

2. Conservatives are adept at self-perpetuation. More on this later in the book when we get to Idaho.

3. "Aggressive" and "defense" might seem like an oxymoron, but it helps if you visualize the 1985 Chicago Bears.

4. That's right, America, we elected a guy—twice—who wasn't even regarded as being the smartest among his own siblings.

5. I said this, presciently, several months before Dick Cheney shot his friend in the face.

6. Which might explain why George Will seemed to genuinely try to fit me into his schedule, although it never worked out. Maybe he needed the cash.

WANTING TO GIVE CONSERVATISM EVERY IMAGINABLE MEDIUM TO infiltrate my consciousness during The Experiment, I set out to watch films I hoped might have the potential to win me over to the right. So I posted a message on Free Republic, a favorite online hangout for people with right-wing beliefs and loads of spare time. The message read:

"I'm looking for a list of the movies most beloved by conservatives, movies that support conservative ideals, films that make a good argument for conservatism. I figured this was the place to go. If you were making such a list, what would be on it?"

The Web denizens inundated me with recommendations. I took their suggestions, added a few of my own, cross-referenced that list against the selection at the local Blockbuster, and, when I needed to unwind after a day's immersion in conservatism, caught a flick.

For the purposes of this book, I present each film to you in four segments: summary, conservative messages, anticonservative messages, and overall persuasiveness score (on a scale of 1 to 100).

### RED DAWN (1984)

**Summary:** In the film, as presented in hasty exposition in the opening minutes, a bad wheat harvest has meant the Soviets are hungry and so, because the Green Party has gained control of West Germany, the Soviets set off on a long-rumored program of global conquest. Since it seems like as good a time as any for communism to take over the world, Cuban and Nicaraguan troops overpower El Salvador, then Honduras and Mexico plunge into chaos, and NATO dissolves (to be fair, it was a really, really bad wheat harvest). In a unique tactical decision, a coalition of Soviet and Cuban forces decides to begin their invasion of the United States in rural Colorado. After paratroopers gun down a high school teacher, some of the students figure they can go ahead and cut class that day. They gather some firearms, ammunition, food, and ski parkas and head for the hills. The students look like what would happen if the editors of *Soldier of Fortune* took over *Tiger Beat,* circa 1984, possibly by force; the movie stars Patrick Swayze, Charlie Sheen, C. Thomas Howell, and Lea Thompson. While the town's citizens are either shot or herded into detention camps, the teens, nicknamed the Wolverines after their high school mascot and led by Swayze (supposedly a recent graduate of the school but looking every one of his thirty-two years), launch a series of guerrilla attacks against the occupying forces. Ultimately, the repeated car bombings and sniper attacks get to be too much for the Soviets and Cubans and they leave. The nation, under the presumed leadership of aging teen Swayze and his handful of surviving compatriots, looks forward to a better tomorrow.

## CONSERVATIVE MESSAGES

- Communists want to come to your town, air-drop in actually, and shoot your teacher.
- A gun is literally pried from the cold dead hands of its proud American gun owner, thus proving claims made on the bumper stickers on vehicles owned by conservatives.
- The Cuban commander orders his Soviet flunky to

fetch the list of people in town who have guns and what types of guns they have. This indicates that gun registration, while it might seem like an acceptable idea to goody-goodies today, is ultimately a bad idea. Because when the Soviet Union overcomes its dissolution, reforms, and really does invade, we don't want a paper trail.

- Hotties like Jennifer Grey and Lea Thompson can be found in root cellars and are willing to run off with gun-toting teens rather than take their chances with communists. So if you want to get with Jennifer Grey and Lea Thompson, toss away that copy of *Das Kapital,* comrade, and pack some dried deer meat instead.

- Therapy and understanding don't do a lot of good in occupied America. Whenever one of the Wolverines gets weepy or freaked out, Swayze urges the offending party to "let it turn to something else," like vengeance or aggression or something.

- People in traditionally liberal Denver are said to be eating rats, sawdust bread, and sometimes one another. Ergo, liberals are savages.

## ANTICONSERVATIVE MESSAGES

- It is impossible, in the context of the current American situation in Iraq, not to think how easy it would be to recast the entire movie with Iraqis as the Wolverines and American troops in the part of the communists. An army from one country invades another country acting on what it says is best for the invaded country. The populace sees tanks, soldiers, and civilian deaths piling up and launches insurgent terrorism-style attacks against the occupying forces. There's even a prison where some very unpleasant things are going on. So when you watch this film today, it creeps you out. And in case you're slow to pick up on all that, a Cuban officer, discussing the Soviet invasion of Afghanistan, says, "Look, I was always on the side of the insurgents."

- Reagan had been president for four years by this point. So whatever he was doing obviously wasn't doing a whole lot of good.

- Any country that depends on Patrick Swayze to save it probably isn't worth saving.

**OVERALL PERSUASIVENESS SCORE: 34**

# 5

# Neoconservatism

### Becomes

## Kristol Klear

☞

**In which the author is unexpectedly forced to come
to terms with a deceased family member while
simultaneously grappling with comprehension of the
dominant foreign policy philosophy of recent years and
that's quite a lot to deal with all at once when you think
about it, I mean, wow, but yeah, there it is.**

After twenty-four hours in New York City, I was on a train
heading back to Washington, D.C., and I was back on the
iPod. Craig Morgan has a song that complains about someone
at an intersection who's talking on the phone instead of driving. Those don't mix, he says. I agreed with that. Was driving
and talking on the phone at the same time a liberal thing? Or a
conservative thing? Or was it just something that is committed
by dickheads on all parts of the political spectrum?

Next up was Lee Greenwood, a favorite of the traditional
Republican infrastructure. When you want to throw a fund-
raising event and invite old wealthy people, you book Lee
Greenwood to sing. He's bland, he's friendly, he's completely
nonthreatening.

But here's the thing about Lee Greenwood: dude can't sing.
He sucks. Lee Greenwood sucks.

Now, I can value anyone's artistic output even if I differ with them philosophically; Richard Wagner wrote some outstanding operas and was rabidly anti-Semitic. My criticism of Greenwood is not based on his politics or even his "Doug Henning at a Bible camp" appearance. It's based on the actual singing. I guess he's, what, a tenor? But when he reaches for a high note or, God forbid, tries to hold a high note, the guy sounds like a donkey being strangled by an electrical cord. So why, I wondered, watching New Jersey fly by, is Lee Greenwood so popular? Because he's bland, harmless, and patriotic in a really soft, generalized way? Are conservatives willing to let go of the notion of artistic quality because some hack squawks that America is neat-o? Really?

Next up on the iPod shuffle option was Charlie Daniels with a song that I had not heard since I was about eleven years old and my friend's dad had it on cassette in his truck: "The Devil Went Down to Georgia." While I accepted it at face value back then, there were some things that bothered me this time out.

### 39 QUESTIONS FOR CHARLIE DANIELS UPON HEARING "THE DEVIL WENT DOWN TO GEORGIA" FOR THE FIRST TIME IN 25 YEARS

1. The Devil won that fiddling contest, right?
2. Because isn't that totally amazing fiddle feedback thing the Devil plays (which sounds like Hendrix gone bluegrass) a hundred times better than that high-school-band piece-of-crap tune Johnny plays?
3. I mean, come on, right?
4. And since the Devil is so clearly better, why does he lay the golden fiddle on the ground at Johnny's feet?
5. What kind of one-sided bet was that anyway, your eternal soul for a fiddle?

6. Shouldn't it have been something like Johnny's soul or the eradication of Evil?

7. Or maybe a golden fiddle against some object Johnny placed great value upon?

8. If the Devil went down to Georgia 'cause he was looking for a soul to steal, why does he arrange what appears to be an honest competition?

9. Was there actually some hidden theft or scam going on here on the part of the Devil?

10. Then why not explain that, Mr. Daniels?

11. And who was judging that contest?

12. Was it an honor-system kind of thing?

13. With the Devil?

14. Honor system with the Devil. How did Johnny get sucked into that one?

15. Johnny isn't the sharpest tool in the shed, is he?

16. Was there some sort of arbitration board in place in the event that the outcome was not obvious?

17. If so, who served on this board?

18. It wasn't the demons, was it?

19. 'Cause even though they're the only characters in the song, they're kind of biased since they're in the Devil's band and they're demons, right?

20. So why—why—does the Devil take the dive and throw the contest?!

21. I mean, the Devil can't be hurting for cash. How much is it going to cost him to buy a new golden fiddle?

22. I'm thinking maybe $18,000. Does that sound right to you?

23. If you're Johnny, what do you even want with a golden fiddle?

24. Doesn't the metallic surface of a golden fiddle create an unpalatably tinny sound as opposed to the nice resonant sound on a wooden instrument?

25. Does Johnny think he's going to display it in his home and tell people the story of how he beat the Devil?

26. Who's going to believe that?

27. Or does he try to sell the fiddle?

28. If so, how does he go about getting something like that appraised?

29. Or does he just melt it all down for the gold?

30. That sounds awfully hard, don't you think?

31. And is Johnny haunted by the question of why the Devil let him win like that?

32. Was there some catch in the contest that Johnny wasn't aware of where the Devil really does get his soul anyway and Johnny didn't notice it because he's not all that smart?

33. And even if he didn't get Johnny's soul, what is Johnny going to say to God in heaven when he has to explain that he bet his soul, the essence of life, God's one true gift, on a fiddle contest?

34. Johnny knows deep down that he's not really the best that's ever been and that's the source of his insecure boasting, right?

35. Was it really necessary or wise to invite the Devil to come on back if he ever wants to try again?

36. 'Cause what does Johnny need, a second golden fiddle?

37. Or maybe a golden viola the next time?

38. Why would the Devil need an invitation?

39. Are you implying, Mr. Daniels, that Johnny actually wants to get hustled?

That resolved, I turned my attention to *The War over Iraq: Saddam's Tyranny and America's Mission,* which I wanted to finish by the time the train rolled into D.C. because I was going to need to take a cab from the train station to the office of the book's coauthor William Kristol, the powerful neoconservative and editor of the magazine *The Weekly Standard*.

One would think that justifying something that had become as unpopular as the Iraq war might take a while, but Kristol and *The New Republic*'s Lawrence Kaplan wrap it all up in a tidy 170 pages. Between the two authors, the book probably took part of an afternoon. Saddam is dangerous, they argue, and so we've got to stop him. Done and done. It's an optimistic little volume, written before the invasion and before the insurgent attacks and the car bombs and the horror, the horror, the horror; a glimpse into the "we will be greeted as liberators" period in military strategic history. The *Weekly Standard* lobby was a step or three above that of *The National Review.* I guess when you're the epicenter of modern neoconservative thought in America you can afford nice chairs. As I sat in one, I read the current issue's cover story, "A Decade of Reed" by Matthew Continetti. It was about former Christian Coalition scion Ralph Reed and how his contacts with Tom DeLay and lobbyist Jack Abramoff were endangering his bid to become lieutenant governor of Georgia. The article emphasized that Reed's interests and passions have always been much more about politics and power than any kind of religious conviction and that those same motivations were now jeopardizing his political aspirations. It was a fascinating and highly critical examination of Reed, made especially compelling by being loaded with dirt on some disreputable and ethically questionable activities in his past. I would have expected to read it in *The Nation.* Any liberal would, of course, be delighted to see Reed taken down, but was it okay for the purposes of The Experiment to be happy about harsh criticism of a conservative icon when it's being leveled by an iconic conservative magazine? Or was this an example of the conservative body killing off one of its own weaker members to consolidate power? Ideological Darwinism? Was Ralph Reed the Fredo Corleone of the right?

William Kristol was not wearing a suit. I was. What was up with this whole dress-code thing? When did the Ruling Elite start dressing like a dot-com on a Friday? Complicating matters even more than apparel was the fact that William Kristol is my dad. Not my real live actual dad, that one died in 1999, but Kristol looked shockingly, distressingly, unnervingly, like my late father. Or more specifically how my father looked when he was about fifty-five years old and how he would look if he was one of the leading opinion makers of the neoconservative movement instead of a Norwegian freight company manager in Seattle. So it was sort of hard to concentrate on what Kristol was saying. Go ahead and try to concentrate when your dad has been raised from the dead and turned into a regular panelist on Fox News.

As we sat down in Kristol's spacious, elegantly appointed office, in chairs that probably cost more than my first car, his eyes were in constant motion around the room, looking everywhere except directly at me. He had a vague smirk on his face.

I explained the premise of The Experiment and how I was looking to be converted. "I don't like the rhetoric of conversion. It makes it seem like an irrational leap of faith," he said.

Okaaaaay. But was it possible to do it? "Of course it's possible because a lot of people have done it, including my parents' generation—en masse, so to speak." This was his first mention of the way things used to be many decades ago. "Usually there's some big event or series of events that help trigger it. Stalin in the late forties. A lot of people stopped being entranced by communism and became liberal anticommunists. But I think *this* is a moment when lots of big things are happening in the world and it's a time for liberals to rethink some of their orthodoxies."

"What are those big things?" I asked.

"9/11. And the sense of the world of the nineties, the world of multilateral cooperation and commerce and ideological extremism being defused and the U.N. playing a larger role, that doesn't look quite as likely to be the pattern of the next few years." In Liberal Land, President George W. Bush's administration is often faulted for continually citing 9/11 as the justification for any number of aggressive military or domestic intelligence initiatives. But talking to Kristol, I remembered how 9/11, in fact, did create a sense that things might not be getting particularly better in regard to everyone getting along and being nice.

Culturally, he said that whereas people used to feel like no-fault divorce was a highly evolved notion, now there's more ambivalence about it, and when parents have to struggle more and more to keep their kids away from material that would have been considered out of bounds twenty years ago, well, "that's not simply progress." "Why does this guy keep bringing up the past?" I wondered. Then I remembered, "Oh yeah. He's a conservative."

He said the "big thing" in economics happened a while ago but was still pertinent. "Most of that change happened in the seventies and eighties: letting the markets work. It's kind of pointless to fight markets. You can modify them, you can help people benefit from them, but government planning is not the way to go. You don't see a lot of liberals wanting seventy percent tax rates again."

"Yeah, you're right," I said, even though I'm sure I could walk down any street in Seattle and find four within a minute.

While he looked like my father, Kristol's own dad was something to be reckoned with as well. Author and political philosopher Irving Kristol is credited with founding what we know as neoconservatism along with Senators Henry

"Scoop" Jackson and Daniel Patrick Moynihan, who were both Democrats.[1]

I asked Bill Kristol whether he had a single moment when he became a conservative. "When I grew up my parents were still moderate liberals," he pointed out. "They were antileft but they voted for Johnson and Humphrey. The first Republican my parents ever voted for for president was Nixon in '72 and I was twenty years old then. And my parents were gradually becoming more conservative, so in that respect it was easy for me to become conservative."

We talked about the gap between liberals and conservatives, each of whom tend to see their own side as absolutely correct across the board and the other as inherently wrong about everything. Kristol said that while having a political inclination is fine, sometimes one side or the other just gets it right on a particular issue and is proven right by history. He says liberals were right about getting the United States involved in World War II when the conservatives of the day thought we should let Europe settle its own affairs. Liberals also proved correct on the civil rights movement with their whole "racism is bad" concept. Kristol feels that a mentality arose out of Vietnam where liberals adopted opposition to military intervention as a default position and that mentality, not present in the minds of Democrats like Harry Truman, has colored the national discourse on matters of foreign policy. Liberals inevitably point to Vietnam as proof that military involvement is bound to be disastrous.

Meanwhile, conservatives have been on a winning streak on major decisions since then. "At the end of the day, Reagan said, you know, what if we follow basic free-market economic policies? We'll revive our economy. We can quibble about details but basically they were right. And if we stand up to the Soviet Union, they'll crumble. And he was right. And I think those two very big judgments being right is extremely important. The liberals said if you elect Reagan we'll have a nuclear war, we'll have a terrible recession, and the country will fall apart. And we didn't."[2]

I noted that all those decisions he listed involved bold actions on the part of the federal government. He sort of agreed. "My kind of conservatism, neoconservatism, is much friendlier to bold government action than traditional conservatism. Most liberals I know are kind of inclined toward a more neoconservative conservatism, which does involve the state being important."

*Neoconservative* is a term that is used exclusively in a pejorative sense on the left. In recent years, liberals have started using *neocon* as a sort of all-purpose put-down, finally retiring the veteran insult *fascist*.[3] It was jarring to hear Kristol call himself a neoconservative even though I had heard him called that a thousand times. Like when people are gay and use the word *queer* to describe themselves. You feel like it should be an insult, but if they're not insulted, what can you say?

He continued. "Basically, neoconservatism accepts the New Deal, it's pro–civil rights, it's pro-assertive American foreign policy, assertive both in terms of our interests and strengths but also in terms of our ideals. And so it is, if you want to put little names on it, Reagan, Giuliani, McCain, Bush to some degree on the foreign policy side. As opposed to on some days Robert Taft or Goldwater. You don't have to define conservatism as much by cutting government programs as getting the economy going. When liberals get mugged by reality they tend to become neoconservatives not paleoconservatives."

Frankly, this was not what I knew about neoconservatism at all. Where was the bloodthirsty war lust? Where was the pile of Halliburton cash? Where was the evil plotting?[4] These were real philosophies, not sinister plots. Of course, I had left my counterarguments at the door and was only listening, not talking back, not shooting down what he was saying, just opening my ears. Neoconservatism, as Kristol explained it to me, accepts the idea of government being a force for good. I can get behind the civil rights thing, certainly. And I think we can all agree on the New Deal being a pretty positive development. Yeah, okay, the neocons do get a little—shall we say pushy?—every now and again in regard to military ambition, but to simply retreat from the world and hide out in a bunker at home seems unrealistic given the way the world works today. And even though it was led by President George W. Bush, the removal of Saddam Hussein from power in Iraq was, in many ways, a liberal mission. The administration, generally regarded as being conservative, took that mission on and so it became a conservative cause but . . .

helping out . . .

people who are oppressed . . .

stopping the oppression . . .

My head was spinning. It sounded to me like neoconservatism was, in the opinion of its proponents, a sort of all-star team of ideology. Pulling from the right and the left to make this new thing that Pat Buchanan and Teddy Kennedy could hate equally. Here was a reasonable man explaining this philosophy in terms that were clearly meant to enhance my understanding.[5] I was being drawn to neoconservatism. Essentially, it was like being a liberal but kicking more ass and it had more power. I was drawn to the power. Mmmmm . . . power.

"One of the strengths of conservatism in the last twenty-five years is that it absorbed the strengths of liberalism," Kristol said, the interview becoming a little more vampiric. "When you stop having the debate is when you risk turning in on yourself and becoming more orthodox and more monolithic. That's why I think it's very important that the conservative movement is very open to liberals coming in. It's what gives you new thought, new blood, new perspective on things."

There was clear logic and some appeal in the brands of conservatism put forth by Kristol and Lowry. Clarity is not always an easy commodity to come by when you live on the left. Much of liberalism is built on the idea of uncertainty. Iraq was opposed on the left because something could go wrong, tax-cut programs are opposed because they could further harm the economy, the Patriot Act is opposed because it could lead to a police state. Not to say that liberals were wrong about any of those things, but they were all at least initially based on speculation. Conservatism begins with the premise that the reasoning and outcome are not matters of conjecture and guesswork but rather are certainties. If A, then B. It's a simpler and more ordered approach to the world. I was seeing the appeal. I was beginning to admire it in the same way I admired people with strong religious convictions. Deeply religious people are convinced that this world is merely a prologue to the next one and that if you do the things that it says in a book and go to church and all that, you'll be rewarded for eternity. What a nice life, I always think, when everything makes sense like that. Now if I could only find a way to open my heart and let Ronald Reagan, or someone like him, be my personal lord and savior.

That night, I curled up with former Reagan speechwriter Peggy Noonan. Well, not her, technically, but one of her books. I still didn't really get what was so terrifying about Hillary Clinton and I wanted to understand. She's ambitious? She rubs people the wrong way? Huh? Fortunately, Peggy Noonan wrote a book about it in 2000 called *The Case Against Hillary Clinton* and I was able to pick up a used copy. I was excited to read it also because Noonan can flat-out write. I've read entire Reagan speeches just for her use of imagery.

Unfortunately, there wasn't much to write about Hillary in

2000, and unfortunately, Noonan wrote a book about her anyway. Noonan complains about how Hillary, as first lady, was in a position to accomplish a great deal in her eight years and underachieved. Noonan fills a huge chunk of the book with an extended passage wherein Hillary attends a party full of Hollywood elites (Ted Turner, Rupert Murdoch, David Geffen, and others) and urges them to do more to protect morality and defend our children, speaking in florid paragraphs that sound exactly like Peggy Noonan's writing. After pages upon pages of this, she writes, "Which is just when I awoke. Because it had all been a dream. I had fallen asleep and my mind made it up . . . Oh, it was sad to wake up! I liked that dream." Peggy Noonan, who, I should again point out, is an outstanding writer, resorted to a corny dream sequence to make her points because she had no other way to fill the pages of her book because Hillary had not done anything yet. Now I knew how *Dallas* fans felt when an entire season was all Pam Ewing's dream. Finally, after only 208 pages,[6] Noonan gives up. The search for what was so horrible about Hillary would not end this night. I turned out the light and tried to get some sleep. I had a big day ahead, including a chance to visit somewhere I didn't expect to be seeing for a while.

---

1. That's right, lefties, some of the architects of neoconservatism had parenthetical *D*s next to their names. Think Lieberman. Only more so.

2. There was a recession, of course, but still; two out of three.

3. Liberals held a small ceremony thanking *fascist* for its years of service. There was a cake and everyone signed a big card.

4. To be fair, the Halliburton cash pile may have been stored off-site away from this office, but you get the idea.

5. Was Kristol spinning me? Repackaging neoconservatism in a way that a liberal would appreciate? That occurred to me. But I couldn't

believe someone of his stature would go to all the hassle of doing all that repackaging for some dude from Seattle. Unless he saw The Suit and was momentarily stunned into thinking I mattered.

6. Thirty pages longer than William Kristol's entire rationale for war in Iraq. Are conservative pundits naturally more succinct? Or do they have more aggressive editors? Or is everything just a lot simpler for them?

# 6

# Ideologically

## Ambiguous Mermaids,

## Family Research,

## and the NBA Finals

☞

**In which the author meets a pundit, doesn't
meet a prominent Republican, discusses some
of the finer points of Louisiana family law, learns
of some gray areas in the categories of coffee and
newspapers, and must decide where Rasheed
Wallace exists, politically speaking.**

If you need to walk a mile in someone's shoes to understand
them, I should have understood conservatives by now since
I had logged many, many miles in their footwear throughout
Washington and New York. So either that axiom is bull crap
or conservatives experience constant searing foot pain and I
understand them perfectly. On day four of The Experiment,
I put 3.6 additional miles on the Top-Siders from the Rosslyn
Metro station, across the Potomac, up a big long hill, through
Georgetown to the posh Palisades neighborhood to meet with
Jonah Goldberg, editor-at-large of *The National Review*'s
online edition. I could not take the clean and efficient Metro
train there because the Metro doesn't travel to this neighbor-

hood. It's assumed that everyone in Palisades has their own car, if not their own driver.

As with Lowry, Kristol, and "Gannon," I had asked Goldberg to try to convert me to conservatism. He was on his way to the airport that morning, the only day we would be able to meet, but was willing to rendezvous at Starbucks for an early-morning chat/indoctrination. I wondered for a second if this was a special alternate parallel-universe Starbucks (where the naked mermaid wore a pert red business suit like Nancy Reagan) because the Starbucks I knew gave all of its political contributions to Democratic candidates. But if one of the right's up-and-coming hotshots wanted to convene there, I would have to go along with it. I had seen his head shot, spotted him, dressed in the same blue-shirt/khaki-pants combo that must be part of a directive from Karl Rove himself, and he greeted me warmly.

Compounding the cognitive dissonance was what was printed on Jonah Goldberg's cup: JONAH GOLDBERG. Not a picture but a quote by him printed as part of the "The Way I See It" series, where famous and semifamous people are quoted on paper cups for customers to think about before discarding a few minutes later. I'm still not sure if that was the cup randomly given to him by the barista or if he travels with a set of them in order to create what appear to be happy self-promotional coincidences. He complained that the little cardboard sleeve that saves your hand from being burned was blocking out his quotation. Weirding me out further was a copy of the allegedly lefty-biased *New York Times* on his table. I was all *Hey what's up there, Abbie Hoffman, how did the vegan breakfast at the protest rally go?*

It seemed hypocritical to me for conservatives to complain about the perceived left-wing bias of the *New York Times* and then to see one of the leading voices in modern conservatism browsing a copy at a Starbucks. As a liberal, I've become conditioned to always look for hypocrisy among conservatives. It's consistently easy to find.[1] But Jonah said that the left obsesses too much about the idea of hypocrisy.

"It's a hang-up. I'm not saying hypocrisy isn't interesting, I'm not saying hypocrisy isn't useful. But if you're a liberal looking to switch sides, I'd say one thing you're going to have to give up is this weird mainline addiction to pointing out hypocrisy. We saw during the Rush Limbaugh thing, who I'm not a huge fan of, but he was condemned mercilessly for saying that drugs are bad while doing drugs. But no one condemned him for doing drugs; they just didn't like the hypocrisy."

So if step one in the conversion process was forgetting about hypocrisy, what was step two? He said I must recognize how modern conservatism is all about defending and preserving a notion of the American Revolution that is actually liberal.

"American conservatism wants to conserve the institutions that keep us free. Talk to any major movement conservative and invariably the things that have animated conservatives over the last fifty years have been the Constitution, civil rights, the order of communities. All these things are the stuffing of what was at the time a remarkably liberal revolution. The American Revolution was based on the best principles of the Enlightenment, without the worst excesses of the French Revolution. So there are all sorts of conservatives in America today who will also say without embarrassment that they're also classical liberals. That doesn't mean that they don't believe in traditions, because, as we all know, the original founding fathers were all classical liberals. I guess there were exceptions, but all the good ones were classical liberals of one stripe or another."

"The *good* ones?" I asked, inside my own head. Between

Kristol's love of the New Deal and Goldberg's fondness for the liberal leanings of the eighteenth century, these conservatives had a lot of lefty love for all but the still living.

"All of them believed in traditions. With a few exceptions, all of them believed in the role of God and community and all these deep conservative beliefs, but they all believed in the power of the individual and the limited power of the state. So if you're talking about ideological conservatism, about the movement of intellectual conservatism in America, it really shouldn't be that much of a switch for you. It should really just be returning to your original principles."

So all I had to do was cotton to the idea of the founding fathers being self-determinists. Sounded easy. But Jonah said that it also would require consistency. "You can't pick and choose. You can't say one right in the Bill of Rights is more important than another right. If you're going to say that the First Amendment is the most important amendment in the world because it's been in the Constitution, you can't just say, well, we don't really need to listen to what the Second Amendment says. It's all or nothing. And if you believe as an American liberal today that you have the right to do whatever you want in your own bedroom, which I basically believe, then you should also have the right to make economic transactions with who you want. And to say that you don't have economic freedom but you have this other sort of libertine freedom of being able to put a crucifix in urine, is to say that you want one leg of the stool and not the other."

I was a little rattled by all that imagery toward the end there, but I got where he was going. Conservatism means you take what the Constitution says, you trust that the answers are all there, and you buy into it. Unlike any philosophical discussion on the left, teeth are left ungnashed, hair unpulled, questions answered. Everyone sleeps fine. I was also pleased to see the reemergence of the stool in Jonah's description. Rich Lowry, Jonah's *National Review* colleague, had employed the same analogy two days earlier and I told Jonah that.

"Did he really? I thought I just made that up."

But if we're all free to put crucifixes in urine and shoot guns and engage in free-market economics, possibly all at the same time, isn't that closer to libertarianism? Yes, he said, and that's where he was guiding me. Apparently I wasn't ready for the Yamaha motorbike of conservatism, so he recommended the moped of libertarianism. "The beauty of libertarianism is you get to be rebellious and opt out of the regime and all the assumptions of your professors and be provocative, while at the same time you don't have to draw any moral judgments about people doing drugs or hot three-ways in the dorm room next door."

Apparently Jonah Goldberg had sent me hurtling back in time fourteen years or so to a college experience that was much more interesting than the one I actually had. I tried to stay with him.

"So it's a perfect mix of cop-out and rebellion. And then you can start putting things back together. You can say, 'What are the other things that conservatives say they disagree with libertarians about?' and then start heaping them on."

If libertarianism didn't fit perfectly, another direction from which to approach conservatism, Jonah said, was a brand of populism that evolves into nationalism. *Populist* is a term I normally associate with Tom Harkin, Howard Dean, Richard Gephardt, and other Democratic presidential aspirants whose hopes wither in the state of Iowa, but Jonah sees populism elsewhere.

"Fox News has some ideologically right-wing things going on, but there aren't tens of millions of who are watching it and really agreeing with free-market tax policy. The reason they're watching is there are a lot of little flags around. A lot of Americans are actually proud of being Americans and they think that Americans are the good guys. What Fox is doing isn't that radical, it's actually a throwback to an earlier form of American journalism. A lot of what passes in the heartland for conservatism in America is really a sort of mild nationalism, and nationalism is normally associated with the right, it shouldn't be but it is, and you're going to have to feel for where you come down on a lot of this stuff.

"If you're going to be a conservative, you're going to have to look in the rearview mirror a lot more often. And you're going to have to say, 'Well, what would my liberal champions have thought about?' Kennedy was a nationalist: 'Ask not what your country can do for you.' The New Deal was drenched with nationalism on the populist side."

These were a lot of marching orders: ignore hypocrisy, endorse every part of the Constitution (I've always been against being forced to quarter soldiers in my house, so that part will be easy), try libertarianism and don't get bothered by the hot three-way action in the dorm room next door, watch Fox News for the mild nationalism, and reflect on my personal relationship with FDR and JFK. Whew. Okay. "How about country music?" I asked. "Do I have to eventually like country music?"

"No."

"Because I've been listening to Lee Greenwood and he sucks, right?"

"I don't even know his stuff well enough to tell you whether he sucks."

Okay, I said. I think I got it. Unless there was anything else. There was.

"Maintain a distinction in your mind between Republicans and conservatives. I have no great pride in being a Republican. None whatsoever. I have great pride in being a conservative. Not that there aren't many, many, many, many things the Republicans do on a day-to-day basis that drive me crazy, that disgust me, that are horrifying to me. Because they're a violation of conservative principles, and you have a lot of Americans who believe Republicans and conservatives are synonymous. They are not."

I told Jonah that this might make the next day kind of interesting for me since I was planning to attend the College Republican National Convention in nearby Arlington, Virginia. He cautioned me to be wary of young people who identify themselves more by party affiliation than by philosophy. "People who call themselves 'college Republicans' are generally about a notch or two dumber than people who call themselves 'college conservatives.' If you're nineteen years old and getting all excited about Elizabeth Dole, there might be something wrong there."

After our interview, Jonah offered to give me a ride back down the hill, a 3.6-mile lift to the Metro train station whence I came. I gladly accepted. He apologized for his late-model and, yep, conservative sedan looking a little beat-up; it had sat outside while he and his wife were on vacation. Turns out he had recently returned from the San Juan Islands off the coast of Washington, where his wife has family and where the two of them had gotten married. Jill and I had first met up in the same islands and had been married there ourselves. Jonah Goldberg and I had so much in common: we were about the same age, we both think about politics a lot, we enjoy Starbucks and the *New York Times,* and we got married (to women, not each other) in the San Juan Islands. We might even have been friends were it not for the geographical separation and the yawning chasm that separates our worldviews.

His conservatism was pretty simple: the country had been founded on good principles and we should try to preserve those as much as possible. And there was no hidden global conquest message in what he was saying.[2] Though I wasn't about to get a Ronald Reagan tattoo, I found the experience of sitting down and talking to Jonah Goldberg (and Rich Lowry and William Kristol and maybe not so much "Jeff Gannon") extraordinarily valuable. They'd all given me a great deal to think about. But these meetings had all been prearranged. Now it was time for a drop-in.

After checking into my downtown Washington hotel, I changed out of the blue polo I had been wearing and into the Puffy America Shirt. It was a scorching-hot day, so I ditched the khaki pants in favor of shorts. And since my only other option was the dress shoes, I stuck with the Top-Siders. Preppy from the waist down, Wal-Mart frequent shopper from the waist up, I hit the streets. I walked a few blocks to 801 G Street, the headquarters of the Family Research Council. According to the organization's Web site, they champion "marriage and family as the foundation of civilization, the seedbed of virtue, and the wellspring of society. FRC shapes public debate and formulates public policy that values human life and upholds the institutions of marriage and the family. Believing that God is the author of life, liberty, and the family, FRC promotes the Judeo-Christian worldview as the basis for a just, free, and stable society. No gays allowed in our clubhouse!" I made up that last sentence.

It was a few days after a *New York Times Magazine* article (that I read on the last day before The Experiment) had profiled the center, its policies, and even its gift shop. Given that exposure, I expected the place to be overrun with fawning righties celebrating the center's virtues and angry lefties looking to pick a fight. Instead, the Family Research Center offices were not merely abandoned but locked. Seeing me wandering around looking for another entrance, a security guard came out and let me in. Apparently I had passed the stringent "guy who seems to want to come in here" security test. Off the lobby, I entered the posh and roomy gift shop, which doubles as a library and reading room. When you don't need to stock a broad cross section of viewpoints, there's plenty of room for comfortable chairs. There were books arranged on clear glass shelves, some from conservative mainstays like Peggy Noonan and George Will and some from employees of the FRC itself.

And man, were they ever upset up about gay people. True, there were books about the benefits of homeschooling along with a screed or two against judicial activism and how to get all the liberal judges fired, but aside from that it was gay gay gay.[3] I've been in gay bookstores run by gay employees in gay neighborhoods that weren't as interested in gay as this place. There were books highlighting how inherently destructive gayness can be to the gays and how the vaunted "coming-out" period of exhilaration always gives way to empty meaningless sex, abandonment, and disease. I made a mental note to tell a friend of mine, who has been with the same partner for some twenty years that he must still be in the exhilaration period and it will soon all fall apart.[4]

There was stuff to buy other than books. Shirts, bright orange baseball caps (which actually looked completely gay), and water bottles (for when you're thirsty AND against the gays), all sporting the FRC logo. There were copies of the Ten Commandments on posters that you could buy. There was a poster about how cool it was to not have sex if you're a teenager and how you could do lots of other hot and hip things instead like go hiking or play the saxophone or, evidently, join the band Cameo. The art direction was very splashy, eighties-style with rad way-out colors, out-of-fashion fashions, and floating blobby shapes. It looked like what would have happened if the gang from *Saved by the Bell* had been hired to create an abstinence poster.[5]

In the reading-room section of the gift shop, they had some nice chairs set up and newspapers on those big wooden rods used at libraries. Weirdly, all the newspapers were only sports sections. *USA Today*'s sports, the sports section of the dreaded liberal *Washington Post,* all sports and no news from lots of sources. This was a tremendous relief to me. The other night in New York, I had read *USA Today*'s sports section even though I wasn't sure where that paper stood on the spectrum. I had felt guilty ever since, but now it was fine. I was relieved and delighted to be able to welcome back sports coverage and Starbucks coffee in the same day. Now I could buy a paper, toss out most of it, and keep the sports section, much in the same way that someone on the Atkins diet throws away the bun on a Big Mac and just eats the meat and cheese.

I asked the guard if there was anyone I could talk to about what the Family Research Council is and what they do. He said sure, made a quick phone call, I read more about the destructive power of gay, and soon a well-dressed woman of about forty, elegantly coiffed in a hairdo that was equal parts Laura Bush, Nancy Reagan, and Darth Vader's helmet, came down to answer my questions. My policy during The Experiment was not to lie about anything, but withholding certain truths (my political heritage, the conceit of The Experiment) was fair game. I told her I was from Seattle and visiting Washington, D.C., and I wanted to know more about the organization. She said the Family Research Council is an advocacy group for family issues. "Like what?" I asked her.

"So anything related to the family," she said.

"Like . . . what?" I asked her.

Right now, she said, they were focused on the marriage issue and campaigning for the constitutional amendment for marriage. "For same-sex marriage," she clarified. "You guys are FOR same-sex marriage? You want to pass an amendment

FOR it?" I asked incredulously. I swiveled my head around and made an astonished face to underscore the drama. I was having a lot of fun.

"No no! We're against it!" she exclaimed. "We believe that marriage is between one man and one woman!" I stopped my fake hyperventilation and asked what other issues they had on their plates.

She said lately they've been doing a lot of work on the stem-cell issue. The FRC is opposed to the use of stem cells gathered from embryos and fetuses that had not come to term, whether through abortion or some other means. They do support adult stem cells, gathered from the marrow of consenting adults (straight ones, presumably). According to their research, the embryonic stem cells had not been successful anyway in curing diseases but the adult ones had shown remarkable results. I wasn't there to judge, I was there to listen.

"Wasn't Nancy Reagan in favor of the embryonic stem cells?" I asked. Ronald Reagan had died of pneumonia brought about by Alzheimer's and Nancy, along with most Americans, thought that stem-cell research might have helped.

"I don't know. I know Ron Reagan, his son, was," she said, pronouncing "Ron" in the same way one might say "goat porn," "but I don't know about Nancy." But I knew about Nancy. The whole world knew about Nancy. She supported stem-cell research. Still, I dropped the topic and asked about the leadership of the organization.

She said the FRC headquarters were built in part by a generous grant from Rich DeVos, the founder of Amway, the organization that gets you in an uncomfortable position with your enterprising and desperate neighbor. Onetime fabulously unsuccessful presidential candidate Gary Bauer served as president and now the organization is run by Tony Perkins, not the dead bisexual actor from *Psycho*, the right-wing former legislator from Louisiana. My guide told me that Perkins had been instrumental in getting the marriage covenant law passed down there that makes it harder for people to get divorces. Citing Scripture that mandates that married couples should stay together, they passed a law where if you want to get divorced in Louisiana, it's going to be a huge pain in the ass. Morality enforced through hassle, essentially. You have to go through counseling and then do lots of paperwork, then more paperwork, then wait awhile. Don't let Mardi Gras make you think Louisiana is liberal. They almost elected David Duke governor.

"And wait, that marriage covenant law is based on Scripture?" I asked. She said yes it was.

"But our laws aren't supposed to be based on the Bible, are they? I mean, that's the separation of church and state, right?" I asked, skating juuuust along the line of arguing back and breaking my rule but still sounding genuinely confused and asking genuine questions. "I mean marriage is a legal institution."

"No, it's not," she said.

"Well, sure it is."

"No, it's not. Marriage is between one man and one woma—"

"I understand that. But if I'm married and my wife dies, then I'm the executor of her estate. If we're not married and she dies, I'm not the executor. That's the law. It's a legal construct. So I just wonder how Scripture got introduced into that."

"Well, it passed in Louisiana. And it's becoming really popular around the country," she offered, employing the rarely used "what's good for Louisiana is good for the nation" rule. I know what they're up to at the Family Research Council: they want families to stay together because they think that will help society. But already in The Experiment I had found different conservative approaches to the same issue: Rich Lowry's idealistic "marriage counseling funded by cuts to corporate farming subsidies" plan vs. the Family Research Council's "hassle you into staying married" approach. With the latter scenario, we had a big aggressive government plan to annoy people sufficiently that they remain in their marriage. Its stool leg #2 (traditional morality) battling stool leg #1 (small government). If only the army could get involved and break the tie.

I thanked her for her time and walked outside. Now seemed as good a time as any for some sightseeing.

I walked over to Capitol Hill. The Capitol building was heavily patrolled both by bored-looking people setting up tents for some party and also by Capitol police with machine guns who look right back at you when you look at them and probably weigh whether to shoot you and if so for how long since, after all, they have been given a machine gun. The lines were long to join a tour, and by the time I got there, the actual members of Congress were leaving for the day anyway.

Heading back to the hotel, I happened to pass the Russell Senate Office Building. It seemed like it was open to anyone and I had no plans so I went in. It dawned on me that maybe Puffy America Shirt, with its Wal-Mart pedigree and soaring eagle artwork, was actually a kind of passport to the halls of conservative power in Washington. Wearing it had gained me entrée to the Family Research Council and the office building housing many members of the Republican-dominated Senate.

But I quickly realized that unless you're armed or obviously homicidal, anyone can walk right into the Senate office buildings and look around. You paid for them after all, so as long as you can make it through the metal detectors you're free to roam. They don't even hassle you in the hallways. I've seen a whole lot of fireworks in my lifetime. I've watched patriotic films (and had more on the calendar for the next month). I've heard zillions of political speeches, read a billion America-themed bumper stickers. I've stood in numerous elementary school classrooms, absentmindedly muttering allegiance to the flag and to the republic for Richie Stans. I've worn Puffy America Shirt, Blood Shirt, and Hat Shirt. But none of those things have made me feel as good about America as the opportunity to walk right into the Russell Senate Office Building for no other reason than because I felt like it. You can't do that at the White House or the Supreme Court. The executive and judicial branches aren't accessible like the legislative, the branch of the people, our representatives. I was there for hours.

Knowing he was the early-money favorite to get the 2008 Republican nomination, I found John McCain's office and stood there for a long time, staring at the nameplate and waiting for the magic to sweep over me. After about twenty-five minutes I actually felt the needle move a little. There seemed to be an energy coming from inside the room where McCain was either hard at work or completely absent. Yes, he is often called every Democrat's favorite Republican because he doesn't march in lockstep with the party and criticizes Bush on occasion, but he is a real conservative, pro-war, antiabortion, the whole thing. Then again, he's also in favor of diplomacy, supports the humane treatment of prisoners of war, and he seems aware that global warming is real and that perhaps we should do something about it before Utah becomes beachfront property. Many on the left have a fondness for McCain because he doesn't seem to be a part of the larger Republican establishment and because in the 2000 South Carolina Republican primary, he was on the receiving end of political maneuvering that could be described as either "tough" or "scuzzy," depending on whether you ask a President George W. Bush loyalist or anyone else. McCain had been doing well in the polls and was coming off a strong showing in New Hampshire, but the Bush campaign, or people operating on its behalf, targeted McCain for a takedown with push polling that suggested McCain had fathered an illegitimate interracial child (McCain has an adopted daughter from Bangladesh). Lefties are all about being on the receiving end of Bush politics, so they tend to have a fondness for the Arizona righty.

Later that night, I stopped in at the hotel lobby sports bar, ordered a Coors Light, and caught the last game of the NBA Finals between the San Antonio Spurs and the Detroit Pistons. I didn't know whom to root for to perpetuate my crusade. San Antonio because Texas is a red state while Michigan went blue? The Spurs' Tim Duncan, who fits the conservative's self-image of confident, dispassionate, and cerebral over the Pistons' Rasheed Wallace, who fits the conservative's image of a liberal, being emotional, petulant, and irrational? Seemed like the Spurs were the obvious choice and I started pulling for them accordingly.

But then I thought about the Spurs roster. Manu Ginobili, their star shooting guard, is from Argentina. They had the Croatian center Rasho Nesterovic on their roster along with a reserve guard from Turkey named Beno Udrih. Duncan himself is from St. Thomas in the Virgin Islands, which is part of America but only just. And their starting point guard, the guy running the show, is a Frenchman—from Paris!—named Tony Parker. There was even a guy with the last name Mohammed on the team! Meanwhile, the Pistons are full of American dudes, located in the city that used to make all the world's cars before other countries started doing it better. Rasheed Wallace, American. Rip Hamilton, American. Chauncey Billups, American. True, they had a Latino point guard named Carlos Arroyo, but they only traded for him when no one else really wanted the job and he worked pretty cheap.

By the time I mapped all this out, the Spurs had won the game and I was awfully tired and somewhat drunk on Coors Lights. So it was off to bed. I had to get up early the next morning and go back to college.

---

**1.** e.g., an administration full of people who avoided fighting in a war goes on to start one.

**2.** Unless he and William Kristol had been frittering away their valuable time concocting a web of deceit to spin around me.

**3.** Titles at the FRC include *Getting It Straight: What the Research Shows About Homosexuality; The Bible, the Church, and Homosexuality; Outrage: How Gay Activists and Liberal Judges Are Trashing Democracy to Redefine Marriage;* and *My God, What If We're All Worked Up About Gay Because We're Secretly Gay Ourselves?!* I kind of invented that last one there.

**4.** Come to think of it, that guy does get really bad allergies.

**5.** Which obviously would have been an outstanding episode.

## INDEPENDENCE DAY (1996)

**Summary:** Evil aliens come from far away and start blowing up everything and trying to kill everyone. Though it's not explicitly stated in the screenplay, it seems likely that they hate our freedom. There's no use negotiating with them. You're either with us or you're with the murderous space aliens. Bill Pullman, as the president, is a former fighter pilot who attempts to blow up the alien spacecraft himself. So it falls to nerdy egghead Jeff Goldblum and charismatic fighter pilot Will Smith to overcome their various love-life issues, understandable trepidation toward space-alien combat, and scenes with an over-the-top Judd Hirsch to save the world, which they do through force of determination and a computer virus. Really. That's how they got the space aliens. With a computer virus. From a Mac. Because apparently the aliens run on Mac. No, I'm not kidding. Yes, a Mac.

## CONSERVATIVE MESSAGES

- Bad creatures, not really humans, are coming from outside our borders to kill us all. Appeasement and negotiations are ineffective strategies.
- Bill Pullman sort of looks like President George W. Bush. Sort of.
- After weighing his options for a while, the president decides to "nuke the bastards."

- Enormous spaceships full of evil aliens exploding all over the place is bitchin'.

## ANTICONSERVATIVE MESSAGES

- Despite his resemblance to President George W. Bush, Pullman behaves differently. He actually flies military airplanes in real combat situations, assembles a large international coalition for something (world saving) that everyone can agree on, pulls back the troops when it's clear that the prevailing military strategy is not working, and fires the secretary of defense, who turns out to be this total dick that everyone hates.
- Will Smith's character's girlfriend is a stripper with a young son. Will Smith's character sleeps over at her house. Everyone's okay with this arrangement.

**OVERALL PERSUASIVENESS SCORE: 55**

# "We Get the Party Started!"

☞

**In which the author attempts to reinvent himself as the kind of college student he never was, or, depending on your point of view, becomes that creepy guy who hangs out at the old college hangouts for way, way too long. He does so in the interest of education. But that's what those creeps always say.**

The next morning, I took a cab to Arlington, Virginia, for day one of the College Republican National Convention, the motto of which, and I am so not kidding, was "We get the party started!" While I remembered the words of Jonah Goldberg about college Republicans being a notch or two dumber than college conservatives, I was trying to keep an open mind. I wanted to give these people some credit and not think that they were necessarily the conflagration of dweebs, dorks, and virgins they appeared to be. My intellectual generosity was hampered, however, by my hangover and the fact that these kids had gone and booked Tom DeLay to kick off the convention. Nine A.M. and I would be listening to Tom DeLay while hungover. Dirty pool. At this point in DeLay's career he had not yet been indicted on any of the various charges leveled against him, though it seemed likely he was going to be. He was also a hell of a fund-raiser and was therefore still very much in the good graces of his party.

I found a seat in the very front row, maybe ten feet from the podium. Before the events kicked off, I realized that I was sitting next to some kids from Texas. The kids generously offered me a "Tom DeLay for U.S. Congress" decal to wear, which I accepted and then wiped under my chair like a booger. After all that DeLay had been through with ethics investigations and potential jail time, I wouldn't expect jubilant celebration from his party. Tolerance, sure. Excuses and justifications, maybe. But nope, these kids were waving DeLay yard signs around as if he were running for not just president but Global Emperor.

I asked the kids how things were going with DeLay and if he was going to be heading to jail like I had heard. One of the kids was interning in DeLay's office for the summer and assured me that thanks to a recent legal opinion he had heard about reimbursing lobbyists for junkets, DeLay might actually avoid prosecution. "Wow! Avoid prosecution?! WOW!" I yelled.

"What's your name?" the intern asked.

"Well, it says *J* right here on my name tag!" I said, not lying. I hadn't bothered to list my full name and the initial made me feel like a secret agent.

"Glad to meet you, J," he said.

DeLay took the stage and the crowd went crazy with cheers and screams. I chose to attempt some catharsis and instead of screaming "Yay!" or "Whoo!" like other attendees, I hollered "AAAAAUUGHH!AAAAAAA!!!YEEEEAAAAWWWGH!!!" with my hands on the sides of my head like in the Munch painting. I think he noticed. I sat down in my seat.

After boasting of job creation, new homes purchased by first-time buyers, and a deficit that is projected to mysteriously get somewhat better eventually somehow, DeLay started ripping into the Democrats. He brought up something that had happened the day before the convention started, while I was touring the Capitol, when Karl Rove gave a speech and said, "Conservatives saw the savagery of 9/11 and the attacks and prepared for war. Liberals saw the savagery of the 9/11 attacks and wanted to prepare indictments and offer therapy and understanding for our attackers."

Some had called that remark slander, DeLay said. "That isn't slander, it's the truth," he declared. And the adoring crowd went nuts. Eventually DeLay stopped talking. Needing air, I left the ballroom to take a walk around the lobby.

It was at this point that I realized there might be some drama going on, beyond the carefully scripted program of establishment speakers. Like the 1976 GOP convention in Kansas City, and unlike every major Republican convention since, there was to be a contest for the top position of the sponsoring organization. The chairmanship of the College Republican National Committee was to be decided in the next three days and it would come down to candidates Paul Gourley and Michael Davidson. I knew nothing about either one of them aside from the fact that Davidson's campaign signs did not feature his name but instead said YOUR CRNC and Gourley's, imaginatively, said GOURLEY (pronounced like an adverb of the guy who got the most votes in the 2000 presidential election). Like any heated political race, there were a lot of enthusiastic campaign volunteers on both sides, but because they were college students, there was even more chanting and stickers and flyers inviting people to tropical-themed parties and shouting of the exclamation *Whoo.*

At first, I thought little of this contest, figuring it was a match between two young dweebs vying for a job that held no relevance to the rest of the world. Not so. The College Republican National Committee, with over 1,300 chapters, is one of the country's largest "527" organizations (tax-exempt political groups that are allowed to raise unlimited funds, named after the section of the tax code that allows them to exist.) It operates with a huge budget, wields enormous influence, and is a kingmaker in modern conservative politics. In 1973, young Karl Rove had chaired the CRNC, gaining power by challenging delegate credentials, allegedly digging through trash cans to find political ammunition against his foes, and winning the support of the Republican National Committee chair George H. W. Bush. Years later, names like Grover Norquist, Jack Abramoff, and Ralph Reed (all connected to Tom DeLay's scandals at the time of The Experiment) would drift through the CRNC leadership positions. If you look at the names of people involved in the leadership of the CRNC through the past thirty years, it's easy to see that the organization is a stamped passport to the corridors of power and a career move that can set you up for a life of power and riches. Either Davidson or Gourley would also be landing a $75,000 salary, entrée to the machinery of Republican politics, and an office in D.C. Beats the hell out of waiting tables out of college.

Once again, I had dressed inappropriately. I went with a dress shirt, tan pants, no tie, and, for reasons I still can't explain, the Top-Siders, but College Republicans wear suits. Full-on suits. Dark suits. They look like they could be federal agents pouring out of a tinted-window SUV to make a big bust or like they could either be attending a state funeral or providing security detail at a state funeral. Many of them brought extra suits to the convention so they'd have multiple suits to choose from. These guys were, like, twenty years old and equipped with larger business wear collections than fifty-five-year-old bankers. I cannot comprehend this even today.

Back inside the ballroom, there was a huge video screen and Senate Majority Leader Bill Frist was delivering a recorded speech to the College Republicans, who looked bored and distracted.

"Republicans know what they stand for, Democrats only know what they stand against," said Frist, who had a pretty good point there, to a smattering of reflexive cheers. Watching him drone on and on, watching the lack of response of the assembled multitude, I realized that Bill Frist would never be president.

Ed Gillespie, former chairman of the Republican National Committee, followed with a pleasant and drab speech about how great Tom DeLay is and how marvelously well positioned the Republicans were for the 2006 midterm elections. Tony Perkins from my new friends at the Family Research Council gave a dull and distracted speech about the dangers of liberal judges. I kept waiting to hear him talk about the dangers presented by gay people but when it became evident that he wasn't getting to it, I went out and got a big liberal cup of Starbucks.

On the escalator on the way back up to the convention, I realized I was standing directly behind Phyllis Schlafly: conservative icon, Goldwater devotee, destroyer of the Equal Rights Amendment, featured speaker at the CRNC. I was maybe eight inches from her. She's tiny. When I was growing up in a house full of strong women, the word *sexist* was used in much the same way conservatives used the word *communist* or how liberals today use the word *neocon*. Mandated by my upbringing to defend the rights of women, here I was with Phyllis Schlafly right in front of me. I didn't say a word. You always think you're going to have precisely the right thing to say to people like that, but it so rarely works out that way.

When she actually reached the podium a while later, Phyllis Schlafly delivered a stem-winder (well, as much stem as an elderly woman can be expected to wind) against what she sees as liberal judges deciding cases in a liberal manner. I may have been hallucinating, but I swear she complained that the problem with judges is that they think they're smarter and wiser than the rest of us. I kind of looked around the room to see if anyone else found this argument as goofy as I did since I always thought "being smarter" was sort of the point of being a judge. But among the dwindling number of people listening to the speech (to the nerds' and virgins' credit, they weren't all fascinated by an eighty-year-old scold), there was only rapt enthusiasm and fervent nodding. "Activist is too mild a word for them, that's why I call them supremacists," she said. By the time she finished, the room was half empty.

There had been plenty of speech making about the danger of terrorism, the even-greater danger of liberals and Democrats, and the greatest danger in the known world: judges who have confidence in their own intelligence. But so far nothing at the convention about the, uh, you know, WAR that we were IN at the MOMENT. Here were hundreds of able-bodied[1] young men and women swearing loyalty to the party that had entered the nation into war. It seemed to me that the convergence of demographics and current events should have at least entered that topic into conversation at some point. Maybe in the hall-ways? The restrooms? No?

In the lobby, it was all about Gourley vs. Davidson, and while Phyllis Schlafly was speaking, things had gotten more heated. Gourley was from South Dakota and the current treasurer of the CRNC. He was touted as having exceptional organizational skills, a somewhat pedestrian choice to list as your candidate's best feature. While his campaign volunteers were enthusiastic, they had an edge to them, like they were hiding something. Texas-born Davidson was an outsider, having been a Republican activist at the not-quite-Republican University of California at Berkeley and state chairman of the College Republicans.

I walked up to Joe, a young Oklahoman holding a Gourley sign and bedecked with a nonconservative lip ring to go with a heartbreaking case of acne, and asked him what was so great about their man. After paying, well, lip service to Gourley's organizational talents, he said that his guy was more conservative than Davidson.

"There's rumors," he confided, "that Davidson supports gay marriage."

"Is that true?!" I asked him with as much horror as I could feign.

"It's just what I heard. He's more of a *California* Republican," said Joe, using an emphasis that seemed to hint that Davidson was either gay or, worse, liberal.

I asked him on a scale of 1 to 100 how conservative Gourley was. Joe told me that with 100 being a "right-wing fascist nut-job," his words, and 1 being "somewhat left of center," Gourley was probably a 65 or 70.

"Only thirty points from fascism?" I asked. Yes, he said, but thirty points is a lot of points. I wondered where I'd land on that scale by the time The Experiment was all over.

Next I found a young woman holding a Davidson sign and told her that Joe over there was saying that her guy wasn't as conservative as their guy was. "Well, that's not true," she said, her feelings visibly wounded. Mike Davidson, she told me, is very conservative and he is against gay marriage.

"But aren't California Republicans a little different from other conservatives?" I asked. "You know, *California* Republicans?" I did not know what I was hinting at.

"Not Mike," she said. "He's conservative all the way."

"Then why would that guy Joe *say* something like that?" I asked. She didn't know. She just stared at Joe and shook her head. I shook my head with her. It seemed like the thing to do. I had forgotten how much fun college was.

Though it was initially the big-name speakers who attracted me to this event, the real action was proving to be away from the podium. Up on the lectern, it was a bunch of traditional righties delivering familiar refrains to a crowd that was steadily dwindling. Reverend Jesse Peterson, a conservative activist who is black and markets himself as "the other Jesse," said that almost 70 percent of black children are born out of wedlock and that most black leaders aren't saying that there's anything wrong with that. "Gosh," I thought, "if that's true, that is a pretty depressing number. If it's true. Is that true? I have no way of finding out if it's true."[2] But before long, Peterson, who is in his midfifties, stoutly built, and more than a little skittish, was off on a screed against what I guess would be "the first Jesse," Jackson that is. "This is not about blacks and whites, it's about good and evil, and Jesse Jackson is an evil man." The crowd, almost universally white, cheered. Allegedly Jackson's son either punched or threatened to punch Peterson once and now Peterson is suing him. Or something? It's going to trial and Peterson wants to win a lot of money off of the Jackson family and he's really excited about that. We were well into personal vindictiveness, having long ago left the moorings of persuasive political dialogue but the anti–Jesse Jackson rhetoric kept the young Republican crowd hooked. I sat in the back thumbing through the various handouts I had received about internships and training programs available at conservative think tanks.

Peterson was followed by David Horowitz, the former Black Panther turned right-wing zealot, now in his sixties and looking like the angriest tenured university professor on campus. He presented, unsurprisingly given his curriculum vitae, inflammatory rhetoric. "The future of the free peoples of the world depends on the Republican Party," he gravely warned, ". . . and ultimately it depends on you." I was starting to feel a bit more like a real college student at this point: bored, sleepy, still a little hungover, wanting to go out and hang with my friends and skip the boring lecture from the old dude. Maybe a little more like Homer Simpson in that episode where he goes to college.

Horowitz, like Schlafly and others over the course of the day, thought that there should be more conservative speakers on college campuses to encourage a diversity of opinion. It's a handy coincidence that they are all, you know, speakers on college campuses who stand to make money off such bookings. But the point remains. A wide range of opinion on campuses is a fine idea. This was one of those ignore-the-hypocrisy kinds of situations. In fact, I would like to take this opportunity to urge college professors to assign this book in all their classes to encourage diversity of opinion. If I happen to make a lot of money off such sales, well, that's the sacrifice I'm willing to make. The anti–John Moe forces have dominated our campuses for far too long.

If you promise free food, college students show up. After the speaker sessions had wrapped for the day (summary: "judges bad, liberals silly"), the Davidson campaign hosted a reception with quesadillas and free drinks down the hall in a conference room. There was loud music coming from rock videos playing on large flat-screen televisions. But oddly, even the videos were in some sense conservative, all hearkening back to a simpler time for pop music. So instead of the latest hits, we were treated to "You Shook Me All Night Long" by AC/DC, "Jump" by Van Halen, and "Fight for Your Right (To Party)" by the highly political and left-wing Beastie Boys.[3] Granted, these videos were implicit endorsements of, respectively, rough sex, suicide, and defiance of parents, but they seemed quaint and, dare I say, traditional when screened in the twenty-first century. The Davidson campaign event was going well, lots of people, lots of networking. I struck up a conversation with a young guy who had worked under Davidson on the California state committee. He was passionate about his man.[4]

After rattling off a sort of personal history/Davidson career trajectory, in which he was welcomed into the fold by Davidson even though he initially supported a rival, he explained what Gourley's big problem was: the appearance of Gourley's name on a fund-raising letter that had been sent out by a direct-mail advertising firm that was working with the CRNC. Attached to the letter was an American flag pin and the letter said that if the pin was returned with $1,000, President Bush would wear that exact pin on his lapel at the Republican national convention.[5] The letter was apparently not even on CRNC letterhead but instead read simply "Republican Headquarters 2004." Gourley and the rest of the CRNC leadership had distanced themselves from the letter as much as possible, claiming that it was sent out without their consent, but still, if you have your signature on something, there's only so much distancing you can do.

Hearing about all this, and digging the free Coors beer and chow, I became something of a Davidson man. Some of the kids at the convention were actual voting delegates, elected by their school chapters to vote on College Republican leadership, but the vast majority were just there to watch and to enjoy the snacks. Not being a delegate, I wouldn't have a vote, but one could not help getting wrapped up in it all. Unlike every other election, it wasn't about the issues. Most attendees agreed on the issues (gays, liberals, abortions are bad; guns, conservative lecturers, no access to abortions are good), so in choosing a candidate, I had the luxury of choosing style over substance. The tension between the Gourley camp and Team Davidson was coming to a boil in advance of that evening's Lee Atwater Gala Leadership Dinner, an event where I finally, after a week of trying, dressed right: The Suit. The dinner was a celebration of all the gala leadership that Atwater, another former chair of the CRNC, had provided to the Republican Party before his death in 1991 at the relatively early age of forty. Atwater was the creative force behind, most notably, the Willie Horton ad that some credit for sinking the campaign of Michael Dukakis in 1988 and ushering in a new era of destructive negative campaigning that is still with us today. He later became chairman of the Republican Party, contracted an inoperable brain tumor, expressed regret for his actions, wrote apologies to Dukakis and others, and dropped dead. That night, we gathered to honor him.

The ballroom was decked out for the occasion. Gentlemen were instructed to wear either a dark suit or rent a tuxedo for the night. Meanwhile, the ladies were evidently instructed to dress like bridesmaids or to squeeze into the old prom dress that they wore two years ago in high school. The successful ones looked like Jenna Bush, the unsuccessful ones like Courtney Love. I sat at a table with a couple of other guys who clearly had also shown up with no one else and, like me, were not part of the powerful movers and shakers within the party's well-funded apparatus. We were soon joined by two more guys and, refreshingly, a girl, all from Mississippi State University.

Eric Hoplin, the love child of the Pillsbury Doughboy and Dick Cheney and also the outgoing chairman of the committee, bragged that the CRNC now had an e-mail list of 100,345 addresses, "and all of them Republicans!" Hoplin welcomed a visitor from France (to isolated boos and hisses from some Bushophiles/Francophobes in the back) named Alexandros Sinka. An elegantly dressed young man who had skipped significantly more meals than many of his American counterparts, he was introduced as the chairman of the European version of College Republicans and offered a European perspective on world affairs. After running through the global empires of the last three thousand years, to the puzzlement of convention attendees who appeared unfamiliar with, say, the Ottomans, Sinka offered a simple message: "You can learn from our failures and you can learn from our successes. And we can learn from your successes and also from your failures." There was audible grumbling from the crowd. The idea of America having failures did not play well, especially coming from a French dude.

After Sinka, it was time for the lifetime achievement award. Receiving it was veteran youth organizer Morton Blackwell, who honored the legacy of Atwater by ripping right into the Democrats in much the same way DeLay had done early in the morning and like Lee Atwater had done before renouncing it and *dying*. "Just remember," he said, "when they [liberals] say 'progressive' what they really mean is 'socialist'!" As DeLay had done earlier in the day, Blackwell brought up Rove's comments about conservatives and liberals and therapy and understanding. Blackwell pointed out that nowhere in Rove's comments had he said that this was directed at Democrats but House Minority Leader Nancy Pelosi and Senate Minority Leader Harry Reid went on to complain about it anyway. "The hen that squawks the loudest is the one who laid the egg," he said. It proved, apparently, that Democrats are liberals, which was a real breakthrough and acknowledged as such by vigorous applause.

I was five days into The Experiment by this point and the giddy excitement of voyaging into the unknown had begun to wear on me. Even Columbus must have occasionally wondered if all the exciting things he was going to see were worth the violence and upheaval that might come with them. I needed to eat some political oranges to ward off ideological scurvy. Particularly wearing were the attacks on liberalism, which, despite my efforts at conversion, were attacks against my beautiful wife, my Sierra Club member son, and almost every person I knew back home. No one likes being attacked, particularly if they've already pledged not to fight back as part of an experiment. I felt like Gandhi. If he was trying to become British.

As the banquet wound down, we spent time at our table discussing some of the cultural/commercial quandaries that people interested in politics find themselves in. Having spent the whole of my own college career without a drop of lefty-boycott-target Coors Beer touching my tongue, I knew a thing or two about boycotts. I dated a girl in college who not only refused to order pizza from Domino's because the pizza chain's owner was against legalized abortion and gave money to pro-life groups, she would also call Domino's after ordering from somewhere else to tell them why she wasn't ordering from them. I always wondered what the sixteen-year-old on the other end of the line made of those calls.

These modern college kids were just as worked up but from the other side of the aisle. The modestly dressed girl at my table from Mississippi said that she never shops at Target ("And that's hard? 'Cause I loooove Target.") because they give money to the United Way, which she says is an umbrella organization giving money to, among many others, "groups that are for the gay." I briefly became fixated on her use of the definite article *the* in relation to the word *gay*. "The Gay," I whispered to myself. With that simple linguistic move, she managed to move 10 percent of the population into the category of anthropological curiosity.

I told her that according to my research, Target had given slightly more money to Republicans than Democrats in the last election cycle. That made her really happy. She announced to the rest of the table that she would soon be returning to shop there.

The talk turned to the political activism of my tablemates in the last election. We went around the table so everyone could talk about what they personally did to help Bush get elected in 2004. One of the Mississippians, an especially smarmy fellow, reminded me of those guys who loved being on the high school football team a little too much, but in his case it was the debate team or possibly student council. Today, he had an intern gig on Capitol Hill and said that he had campaigned for Nader in order to beat Kerry. I had heard of this happening, but it had always sounded so upside down, I doubted that it actually occurred. I asked him why he didn't simply go work for Bush if that's who he was supporting.

"Because the better Nader does, the better Bush does," he said, a point that Republicans and Democrats could all agree on.

"But if you liked Bush, why not just work for him? He's the guy you want."

"People on the left aren't going to vote for Bush, but they might vote for Nader and that would cost Kerry. I raised all sorts of money for Nader," he said.

"But is that honest?" I pressed. He had no answer. He had never considered it.

In the unexpectedly lengthy silence, there was a ripple of tension at the table, but fortunately, probably for everyone involved, a member of the Louisiana delegation stopped by to suck up to the Mississippians. After vague mentions of working closer together, the dude left, everyone agreed he was kind of a slimeball, and the Nader tension was defused.

Following the dinner, a debate was held between the charismatic Davidson and the crash dummy–like Gourley, my first chance to actually hear either one speak. The ballroom was flooded with supporters of both as yard signs were waved and chants went up. Davidson was a marvelous talker, quick-witted, passionate, slightly self-deprecating, and possessed of a Clintonesque ability to work the crowd without letting the crowd know it was being worked. He was asked, "Do you consider yourself a conservative or a moderate on these issues: life, taxes, gay marriage, security? Which of the issues is most important to you and why?"

He responded, "Conservative, conservative, conservative, conservative. It's been a lot of fun defending each one of those conservative principles at the People's Republic of Berkeley, let me tell you. As to which one is more important—I think that's a very difficult question to answer because all those are defining issues, especially if you're a conservative. However, I would say that the most defining issue is life. Whether it is in protecting the life of the unborn, or protecting life through its procreation through the sanctity of marriage, or whether it's protecting life through securing our nation, or whether it's protecting lives by having a free life through low taxes. I am committed to these principles. I was committed to them as a Texas conservative when I started. I was committed to them when I was at Berkeley. We had a very good time. I've been spat on, kicked, I lost a date over being a conservative."

Gourley, on the other hand, was like Bob Dole without the spellbinding charisma. He was a refrigerator box, a display of hammers, an old brown shoe. Here he is addressing Davidson and talking about the College Republicans' "field representative" program: "I'm a two-time field representative. I've run youth efforts in New Jersey . . . I've trained the last three years of field reps to be prepared to go out across this country and recruit new College Republicans. I'm ready to lead this field program. I'm ready to lead this committee. It's what I do best. It's what I've been doing. I'm ready to start, day one."

People were laughing at him. He was often interrupted with shouts of "lapel pin!" from the increasingly liquored-up multitude. In the debate, Davidson pressed for a revamping of the College Republican system, creation of regional-political-director positions, and a more aggressive effort to put state organizations ahead of the national effort. Gourley, part of the outgoing leadership, said he doubted that would work and that things should keep going as they were. For an organization that had gone through so much controversy, you would think a "stay the course" strategy would not strike folks as a great idea, but these were Republicans and Republicans often think staying the course is a crackerjack plan.

Davidson was the vastly superior candidate. Also, although he was a conservative, he was, for the purposes of this election, the liberal. He was an outsider, wanted to shake things up, make changes to the system. Gourley was a conservative in the tradition of the first President Bush, keeping things the same simply because he could not imagine changing them.

After the debates, both camps threw parties in separate smaller ballrooms at the hotel. I went to the Davidson shindig, which featured a live DJ, awkward but exuberant dancing, the kind you would expect from young people who owned this many business suits, and so many people you couldn't move around the room. It was pledge night at the popular fraternity and billed as a "Ronald Reagan Eighties Party." Given the expense they went through, it must certainly have been driving the organizers into a staggering deficit that future generations would be stuck paying for. The DJ was playing Bruce Springsteen's "Born in the U.S.A.," the irony of which still eludes Republicans some twenty-two years later. I grabbed a Budweiser and soaked up the giddy enthusiasm for a while.

On my way back to the room, I poked my head in at the very, very different Gourley party. It was a "party" in the same way that when the teacher brings in orange drink and plain potato chips at the end of second grade, they call that a party. The lights were all on, a few people fluttered about trying to keep up appearances, but it looked more like a funeral that people had come to right after a wedding reception. Like Gourley had choked on wedding cake and dropped as dead as his campaign.

1. Some drawn and pasty, many obese, but most able-bodied.
2. It's true.
3. I briefly considered adding these songs to the iPod. But since their inclusion at these festivities was more based on musical squareness than political philosophy, I had to abstain.
4. I told him that the Gourley camp was saying that his man wasn't a real conservative. He bristled. "No, that's not true at all. He's very much a conservative. He's very pro-life." A tip for liberals: you can have a great deal of fun with conservatives by calling their conservatism into question. You might as well be saying, "The word on the street is that you don't REALLY have genitals."
5. What if it had been true? And what if it had all worked out? President George W. Bush standing up there decked out in thousands of tiny metal flags, forming a shimmering patriotic exoskeleton? That would actually be pretty cool.

## DIRTY HARRY (1971)

**Summary:** There's a deranged murderer on the loose in San Francisco and the city government, full of ineffectual liberal bureaucrats, isn't doing enough to stop him. So it falls to Harry Callahan, Renegade Cop Who Plays By His Own Rules, to take justice into his own hands. Harry even manages to apprehend the killer (who calls himself "Scorpio" because it's 1971 and that doesn't sound corny yet), but thanks to a court ruling that Harry's methods were illegal, the killer is sprung. Then, as bad guys usually do when allowed out of jail, Scorpio robs an old man and hijacks a school bus full of kids. Harry gets the bad guy and convinces him to renounce his delinquent ways and get his GED while in prison and express himself through poetry. No, I'm just kidding. Harry shoots him to death.

### CONSERVATIVE MESSAGES

- Criminals are evil. There's no rationalization or bad childhoods or liberal back story here, they're just crazy evil creeps who need to be dealt with. And by "dealt with," I mean "shot."
- Scorpio is so bad he shoots a "Jesus Saves" sign and laughs while doing it. This is what happens when you lose touch with your spiritual foundations. You become a killer. That's the main reason why regular working Americans don't vote for murderous psychopaths on Election Day. Well, that and the fact that they don't do well in one-on-one debate situations and have trouble with fund-raising.

- A bus advertisement tells us that a new Ford Maverick costs about $2,700. That's not really a conservative strength, I just thought it was interesting.
- The hero's tasks, and indeed the movie itself, would be nearly impossible if Harry had to rely on diplomacy and trade restrictions to capture Scorpio.
- Operating outside the big government, criminal-coddling liberal bureaucracy is not only an efficient way of nabbing baddies, it's downright fun. Harry frequently cracks jokes while shooting guys. He's having a ball.
- Harry gets shot in the leg but won't allow the ER doctor to cut the pants off because they cost $29.50. He's no big spender!
- Harry is sent to talk down a suicide jumper. He deals with the situation by punching the suicidal man in the face. Violence solves EVERYTHING.
- Scorpio gets himself beaten up so he can claim Harry did it. Just like all those crafty bastards at Abu Ghraib.

## ANTICONSERVATIVE MESSAGES

- Not really the movie's fault but you can't help thinking about some of Clint's later films, like *Unforgiven* and *Million Dollar Baby*, where he's still poker-faced but bleeding heart and sensitive. Those were different characters than in this film, but it makes you imagine some renunciation of Harry's methodology has taken place within Clint Eastwood himself.

**OVERALL PERSUASIVENESS SCORE: 85**

# 8

# These Crazy College Kids

## Today with Their Nutty,

# Fun-Loving, Ethically Shady

## Backroom

# Maneuvering

☞

**In which the author learns a lesson in how politics really actually work and finds twists in the plot he's been following. Not like twists in *The Sixth Sense* or *The Crying Game* where everyone's actually dead or the chick turns out to be a dude, but intriguing twists nonetheless.**

The second morning of the convention was dominated by the roll call of delegates voting on who would be chairman. Unlike recent grown-up Republican conventions, it was outwardly contentious. Stepping to the floor, none of the assembled parties really knew what was going to happen. Delegates had unofficially committed to both Gourley and Davidson, switched over to the other side, made promises, and backed out of promises. There was wide consensus that Davidson was the hot candidate and the delegates Gourley had still retained were sticking with him mostly because they had promised to do so a long time ago and wanted to keep their word. Gourley was the Gerald Ford of the College Republicans. But Davidson was like Ronald Reagan, charging out of California, making everyone feel good about the world, twinkling. There was a lot of pressure to get the election decided also because, again unlike the grown-up Republican convention, there was a wedding reception scheduled for the room at 3 P.M.

Tempers were hot. According to Franklin Foer, writing in *The New Republic,* a letter had arrived via fax announcing that the chairman of the Missouri delegation, Will Dreiling, had dismissed all his delegates, all Davidson supporters, and replaced them with new delegates who had been flown to D.C., all of whom were Gourley supporters. This could not be verified since Dreiling was not at the convention, having cracked under pressure from various Republican operatives trying to strong-arm his vote and running off to hide out in Nebraska (again, not a widely executed move among adult Republicans). Finally reached by phone, with Davidson himself passing the phone around to various delegations, Dreiling denounced the letter and called it a fake. The original delegates were allowed to remain.

In the run-up to the convention, Patrick McHenry, a twenty-nine-year-old freshman congressman from North Carolina, had been calling delegates from his state and threatening that if they voted for Davidson, they would have no future in the Republican Party. It might seem pathetic that someone that comparatively old would still be hanging around the college crowd like Will Ferrell in *Old School.* But since the CRNC has that designation as a "527" political fund-raising organization, they can raise as much money as they want. In the aftermath of the McCain-Feingold bill, 527s are hugely important and the CRNC is one of the largest. A former member of the CRNC leadership himself, McHenry had leveraged the CRNC's clout to eke out the narrow victory that landed him in Congress. It's quite a system.

Despite the heated and often personal nature of the disputes, those participating in the election of new committee leadership continued to stringently follow *Robert's Rules of Order*. It was like the British Parliament combined with a frat house. There were motions, countermotions, procedural votes, all offered with evangelical fervency. I could not imagine Democrats on college campuses ever getting this far into something like this. They would have long ago broken into smaller groups to talk about their feelings and do some brainstorming where there were "no bad ideas" and then thrown a rally, listened to some Phish records, and smoked pot.

Not included in the laws of procedure is the right of either side to show videotapes. This was important because, again according to Franklin Foer, the Virginia delegation was apparently up to no good and one of their members, the whimsically named Amber VerValin, had fessed up to it on tape. She claimed that under pressure from her state chair, she helped forge documents creating College Republican chapters where none existed. This meant more delegates for Virginia and more votes for Gourley. Despite having this taped confession, Davidson's camp was unable to show the tape on the floor, leaving Davidson to attempt to describe the allegations. A vote on whether to dismiss the delegates in question failed narrowly. Gourley was seeing things break his way from his spot at the front of the room where his close friend Hoplin was overseeing the proceedings. As his prospects looked better, swing voters, not wanting to place their own futures in jeopardy, began to move toward his side.

In a vote of 79 to 94, Gourley won. He won. They voted for him. Paul Gourley became chairman.

I had misunderstood the entire race. Because of course Gourley won. He had to win. The shenanigans had helped him considerably, of course, but it was not enough to swing the entire election. Gourley won because he was already part of the powerful machine connected to the ruling establishment. He offered more of the same. He offered stability and predictability even if it was offered in a kind of fumbling way. He was tainted with scandal but his loyalists disregarded that history. And he won. The Californian Ronald Reagan may have beaten Jimmy Carter, but he lost to Midwesterner Gerald Ford. The entire election process, like the debate the night before, had been overseen by the soon-to-be former chairman, Eric Hoplin. Even though only Gourley's name was signed to the controversial letter, the company sending out the infamous "lapel pin" letter had been paid by the CRNC, of which Hoplin was the chair.

During the Gourley-Davidson election, Hoplin followed parliamentary procedure but was still in a position of guiding the proceedings and he was a Gourley supporter. Besides administering the proceedings (and sitting at a long table at the front, Supreme Court style), each of the five-member CRNC executive council was also a delegate at the convention, including Hoplin, the chair, and Gourley, the treasurer and candidate. All but one of them voted for their coworker of two years and Gourley was elected 94 to 79. It's a system built to reward the ruling power.

The deck was stacked against Davidson from the beginning. Davidson gave a short, pointed concession speech in which he said, "Don't forget about integrity. Most of you have it. Some of you have lost it. But no matter what, have it. If you don't have your integrity, you have nothing." Gourley sat there looking smug, relaxed, and peaceful.

After the dust had settled, there was nothing much scheduled that day. I stopped in at Starbucks, picked up the *Washington Post* sports section, and read that President George W. Bush and Secretary of State Condoleezza Rice had attended the previous night's Washington Nationals game. If the Nationals are good enough for them, they were good enough for me. I saw there was a game in an hour and caught a Metro train out to RFK Stadium.

Once I got there, I realized that I ought to try to be a Washington Nationals fan. The team has been in existence since pretty much April of 2005, after having spent the first thirty-five years of its life in Montreal (French-Canadian!) as the Montreal Expos. So too had I spent the first thirty-six years of my life in a place dominated by a frame of mind that had become an ideological hinterland, not widely understood by anyone in the actual corridors of power. Despite the Nationals' extraordinarily brief history, the fans on this humid Saturday night were going crazy for the team, welcoming the Nationals/conservatism into their hearts/personal ideologies.

When I saw the Nationals/conservatism play in Washington, they were in the midst of a long winning streak, having picked up victories in numerous recent games/elections. It was a far cry from how things were in their old city/construct of Montreal/liberalism, where their attendance/candidates were always down and things sucked/sucked.

On top of all that, the team's primary color is red, as in the type of state shown on maps to go Republican, and they have a huge *W* on the front of their hats. Maybe if I fell in love with the Nationals, everything else would fall into place. I cheered boisterously for my new team, rooting them on to victory in their battle against a foreign power (the Toronto Blue Jays). The Nationals conquered their foes 5 to 2, using an awesome "shock and awe" arsenal of superior firepower. I hoped that the Blue Jays would eventually greet my team as liberators. The Nationals didn't ask for this game to be played, but if the Blue Jays were going to ignore U.N. resolutions it had to be done.

After, I hopped the Metro train back to the hotel, wishing

that politics were as simple and tidy as a baseball game. In real-life politics, or at least the children's version of real-life politics that I had been following for the last few days, the game was played by rules that were not the same for all participants. It was like Gourley had three extra players on the field and was permitted two extra outs per inning. In baseball, a big error will often cost you a game. In politics, a lapel-pin scandal, or even launching a war, will not necessarily block an election.

# 9

# Alphonso Jackson

## Lives

# in Fear

☞

**In which the author finds himself in the tenuous position
of grappling with the implications of religious figures,
cabinet secretaries, the offspring of cabinet secretaries,
all during breakfast. Not only that, it's one of those lousy
breakfasts like when you go to a breakfast event and the
worst thing about the whole affair is the real, actual
breakfast they serve. What's up with that, anyway?**

Jim Saxton, a Republican House member from New Jersey
was on *Fox and Friends,* which I flipped on immediately
after waking up in the hotel room. I had barely slept thanks
to the all-night parties of jubilant young Republicans, many of
who had gone ahead and slightly loosened their neckties during
the wild reverie. Saxton was talking about the war in Iraq, saying that the people being held at the Guantánamo Bay prison
facilities, many without being charged with anything, without
legal counsel, and held for years at a time, aren't accused of
ordinary crimes, they are accused of vicious horrible crimes.
They're trying to destroy the American system of government
and they will be treated as such, he said. And also fairly, he
added as an afterthought.

The host on Fox brought up a deadly attack that day in Mosul where thirty-six people had been killed and two more attacks the day before that killed many more, including several Iraqi police officers. Despite these and several other recent attacks, which only seemed to be increasing, Vice President Dick Cheney had recently said that he thought this was actually a good sign, an indication that the insurgency was in its "last throes," getting increasingly desperate and therefore near its end. Iraq, he said, would be an "enormous success story." It's like watching your house burn down and saying that at least the fire's almost out.

I turned off the television and went downstairs to the prayer breakfast, the final event of the convention. Howard Dean had recently charged that the Republican Party was the party of white Christians. That remark got lots of people upset who said no, it's a party for everyone and that no matter what you believe, you'll be welcome within the loving arms of the GOP. I wanted to see if Howard Dean was right or wrong. So I approached Rachel, a primly dressed young woman who I was told was going to be leading the prayer. I asked her what kind of prayer it was.

She said that it was nondenominational and that she was planning to talk about God, but it was for everyone. "So if you're Jewish, it will be okay?" I asked her. She assured me that it was. "And if you're, like, Buddhist or Muslim or something else, there will be something in it for you?"

Rachel pointed out that she was absolutely not going to mention Jesus by name, but that most people coming to something described as a prayer breakfast were probably going to expect something Christian. She also said they had a Muslim interning in their offices last summer and that there are even atheists in the party. No one, she said, would feel left out by her prayer. "See you later, J!" she said, having read my name tag. Since I had no friends in the party, atheist or otherwise, I located a table full of other people who were there by themselves. Jayson was a casually dressed youngster from UNLV and stood out in a crowd of people who hated the liberal media so much that they had probably never read the *New York Times*. But Jayson had the full Sunday version of the *Times* and the *Washington Post* on the table, poring over coverage of the war. We talked about Cheney's comments about the last throes and the situation over there as a whole. Jayson favored the Iraq position of the first President Bush, which, these days in the conservative movement, is like the being a big fan of MC Hammer. Jayson also thought that we should have snuck in there, assassinated Saddam, and then waited for the country to transform into a sort of "ding dong the witch is dead" type of democracy.

A Canadian(!) girl attending the conference sat down at our table. The two most popular themes for women's apparel at the convention had been "sorority girl looking to marry into the Republican elite" and "girl rejected by sororities and the campus social scene in general, choosing instead to take up with the anti-campus-establishment Republicans." All girls in the former category had staked out boyfriends and were sitting with them. This girl from Canada was of the latter category, so she was with us. She challenged Jayson's sneak-in-and-assassinate-Saddam plan since Saddam's sons had been well positioned to take control of the country and continue their father's form of government. "Well then, we kill them too. Look," he admitted, "I never said it was going to be easy."

They batted ideas back and forth for a little while and I realized that they were actually talking about the war! And disagreeing! On something connected to the President George W. Bush administration! Disagreement on matters of policy was something I had not encountered for the previous several days. People debated Gourley vs. Davidson, but there was either uniformity of thought or simple obliviousness to things like the, you know, WAR WE WERE IN. Maybe that's because they were all of military age but at a big fun boozy convention instead of being blown up in Mosul. But Jayson and Ms. Canada were going at it. It was invigorating!

Then it got frightening. "What do you think of the war, J?" he asked. I had never even considered what I would say in a situation like that, but I also didn't want to come across as hesitant and the rules of The Experiment stated that I must not lie. "Well, it's a mess over there, of course," I offered with a confident tone. "I guess I'm here at this convention," I continued, "to see if I can hear some discussion on what we do there now. It's all fine to talk about the reasons for going over there, but I don't hear anyone talking about what we do now, two years after the invasion, when the war is still going on and a lot of people are dying. Total withdrawal has problems, as does doing what we've been doing."

Another kid at the table joined in. "It's the borders in Iraq that are the problem; people are pouring in from all over the world. We should move the soldiers out to the borders instead of in the middle of the country."

"But then the country will collapse and you'll need twice as many soldiers covering the borders. Where are you going to get them?" He didn't know. "And besides, most of the insurgency is Iraqi people who just want us to get the hell out of their country." I managed to put on a poker face because my actual face would belie that I knew I had gone too far. As I got nervous looks from the youngsters at my table, I reached for my coffee. "So you go to college out there in Seattle?" the border kid asked. The world-weariness, combined with the receding hairline and graying sideburns, had clearly hindered my attempts to blend.

"Well, I live in Seattle. There sure are colleges out there. I like to . . . try to . . . learn things." Three short sentences, all true. By this point, the whole table was looking at me. "I will soon be gunned down," I thought. But then it was time for Rachel to deliver her prayer. I was saved!

Rachel asked us all to bow our heads, which most did except for the visiting Europeans and me. She then led a simple prayer asking for guidance and wanting a sort of force field put around the troops in Iraq so they wouldn't get hurt. "And guide us into the future, Jesus."

Later, at a break in the program, I saw Rachel standing in line for the tiny dry pastries that constituted the actual "breakfast" part of the prayer breakfast. "So Jesus made it in there after all, huh?" I asked.

"Yeah, He just kind of slipped in," she said. Jesus as cat burglar.

Next up was a keynote address by Secretary of Housing and Urban Development Alphonso Jackson, a longtime friend of President Bush from back in Texas. Jackson, stout, bespectacled, and flanked by a cadre of grave-looking security guards who kept looking right at me, covered a lot of the issues that are really closely linked to housing and urban development, like how marriage is between a man and a woman. Jackson said that he has no problem with people living their lives however they wanted as long as they didn't try to push that belief on him. More applause.[1] I really honestly tried to get his point, I squinted hard and really thought about it, but I simply did not understand. Was someone trying to get gay married to Secretary Jackson and he had to fend them off? Were aggressive gay suitors breaking into his house and slipping gay wedding rings on his fingers while he slept?

Regarding the comparison to the civil rights movement

of the fifties and sixties, Jackson, who is African-American, laughed it off, saying when he walks into a room, he can SEE who's black. More laughter and applause. Huh? Jackson said that the Republicans were the party that freed the slaves and led the civil rights movement and continues to do more for black and Hispanic people than the Democrats ever did. Jackson then called George W. Bush the greatest president America has ever had to the barest smattering of confused applause.

Jackson said that when his daughter was in college she was a Democrat. But now that she's had kids and started making money, she's become a Republican "because it's in her own best interests." People cheered loudly for that one.

When Jackson's speech was over, outgoing College Republican chairman Eric Hoplin turned over the floor for questions. I wanted to give this whole enterprise one more shot by turning away from rhetoric and toward policy. I asked what goals and initiatives HUD was working on over the next few years. Jackson turned very serious and listed several projects, including a more aggressive policy of funding the downtown cores of major cities. He explained that more and more people are fleeing the cities, choosing instead to live in suburbs and exurbs, but that without a strong residential and business presence in the urban areas themselves, the system won't hold together. America, he said, was built on cities. It's our source of strength. Jackson's smirk faded, he stopped playing to the crowd and spoke his mind. Suddenly, for the first time in his speech, and for the first time since the convention started really, I was totally down with what a conservative Republican was saying. Helping the cities may be a liberal notion with government becoming more assertive and using taxes to help people through new programs, but it's also conservative, since through working to maintain something that is important to everyone, we conserve a traditional America institution. Meanwhile, strong cities promote a stronger economy that can ultimately operate without aggressive regulation.

What was the conservative party line? What did Republicans stand for? I was still puzzled, but the downtown talk was good enough for me. I got up and left the convention on what would have to serve as a high note.

At the airport, I added some new music to the iPod after carefully combing through hundreds of songs in my alluring-but-off-limits iTunes program to find something else conservative to lighten up the Clint Black–Daryl Worley–Michael W. Smith–Charlie Daniels–Lee Greenwood–Kid Rock rut that I was in. I got a little more, um, liberal with the rules, loading up songs by Lyle Lovett (played at the inaugural), Garth Brooks (heard somewhere that he was conservative), and that "Mmmbop" song my kids love by Hanson (homeschooled).

I was eager to get home to see my family and promote some family values with them. Jill was holding up well considering she had been left alone for a week with two very young and very active children and considering that a case of pink eye had taken hold of the Moe house, leaving everyone itchy, irritable, and quarantined with one another. I wondered if the entire weeklong absence was, in fact, a liberal or conservative thing to do. On the one hand, I was working and thereby contributing something to society, putting in honest labor to help produce a product (you're holding it) that would ideally create a benefit for both me and the consumer. On the other hand, I was neglecting my family and their puffy eyes for my own personal gain. On the first hand, the family would benefit by my efforts. On the other hand again, I had the benefit of being able to blithely ponder these factors while Jill was surrounded by stacks of dirty dishes, discarded toys, and screaming children demanding apple juice while they punch each other in the neck. I guess it's conservative for a wife to stay cheerfully, obediently at home, tending to the children while her man went out to bring home the bacon. But Jill never signed on to be a conservative, so she tends to think that whole model is pretty stupid. It was probably best that I head home.

Some of the country music was growing on me a bit. Daryl Worley has some nice moments and Craig Morgan is actually pretty good. Toby Keith's "Courtesy of the Red, White, and Blue (The Angry American)" is a song that was heavily criticized when it was released because of the truckload of jingoism it contained. Sure that's there, but Keith's song really appears to be about striking back at the people behind the 9/11 attacks, and on that level, liberals can agree with it just fine. "Courtesy of the Red, White, and Blue" was recorded well before this whole Iraq business got started and it doesn't appear that Keith endorses the broader argument that in a post-9/11 world, we all need to fight against all possible threats, like Iraq. Instead he says, in essence, "We got hurt, it sucks, it makes me mad, we should do something about it." So as I sat in the airport listening to it, I thought, I can go along with Toby Keith on this one. I was 25 percent of the way through the time line for The Experiment and that felt like progress. I had followed the lines of thought of Lowry, Kristol, and Goldberg. I had become enamored of the candidacy of Michael Davidson. I had appreciated Alphonso Jackson's urban investment plan (but not his selfish daughter). Now I dug Toby Keith's most controversial song. Perhaps I was on my way.

Michael W. Smith was a more difficult case. Maybe I'm possessed by the spirit of Satan[2] and can't appreciate good Christian music, but I could not get into his songs. Most of the music on the album is bland and unobtrusive, and when they appeared on the iPod's shuffle setting the songs went by without my even noticing most of the time. But some of my darkest moments were during Smith's song "Healing Rain." It's a slow ballad with a little march-tempo thing going on as Smith sings about, I don't know, Jesus or Armageddon or something. That part's fine. The part that's not fine is at about the 3:15 mark (or is it the 3:16 mark?) (that couldn't possibly be intentional, could it?) (oh my God, I bet it is) where the music turns all rock 'n' roll and Smith starts belting it like he's trying to get to Hollywood on *American Idol.* How could God love us and still permit such music? "I'm not afraid!" he hollers. "I'm not ashamed!" Really, Michael? Because I am both of those things when I listen to you.

---

1. I know a lot of gay people and let me assure you, College Republicans, you're not really their type. Between your head start on obesity and your pale complexion, you're no Heath Ledger, if you know what I'm saying.

2. Which would explain my fondness for Led Zeppelin's more arcane albums.

## THE PATRIOT (2000)

**Summary:** Benjamin Martin is a veteran of the French-Indian War, dealing with horrible memories of his brutal actions in battle and trying to live the simple life of a South Carolina farmer. When the Revolutionary War starts heating up, Martin wants no part of the fighting but his son Gabriel does, signing up almost immediately. But when the British burn down his house and kill one of his other sons, Benjamin Martin decides to fight. Leading a ragtag band of citizen soldiers, he employs unorthodox methods to battle the British and help turn the tide for the Americans.

### CONSERVATIVE MESSAGES

- If someone had acquired the film rights to the Second Amendment of the Constitution, this is the movie that would eventually result.
- Benjamin Martin stands up for both states rights and against taxes, saying, "If you mean by 'patriot,' am I angry about taxation without representation? Why, yes, I am. Should the American colonies govern themselves independently? I believe they can and they should."
- There's at least one passing reference to what stuffy impossible jerks the French are.
- Gabriel at one point says to "stay the course," a popular phrase during the Reagan administration.

- Martin literally attacks the enemy with an American flag. Really. A real flag. It's pretty cool, actually.
- Martin sputters about trying to negotiate out of military action, saying, "There are alternatives to war. We take our case to the king and plead with him." But ultimately, a strong and aggressive military (or in this case armed militia) is the only way to deal with the threat of the British.

## ANTICONSERVATIVE MESSAGES

- As in *Red Dawn,* it's kind of hard to shake the whole insurgency-vs.-occupying-army subtext, with the American protagonists being the insurgents and the modern-day audience feeling a bit troubled about that.
- The "fight when you're attacked" rationale works really well seen through events like invading Afghanistan. Less so for Iraq. It would be like Mel Gibson fighting the British by launching an invasion of Portugal.

### OVERALL PERSUASIVENESS SCORE: 76

# 10

# Don't **Have** Gun,

## Will Travel

# Short Distances

☞

**In which the author takes an experiential route
toward exploring the meaning of the Second
Amendment and weighs whether the militia he could
potentially form with a guy named Larry would
be considered "well regulated."**

Inspired by several of the shoot-'em-up films I had been watching, I decided the next day would be about guns. The NRA always endorsed the conservative candidate in any election, and when liberals like John Kerry hold firearms, as he did in some unfortunate 2004 campaign photo ops, it's a generally uncomfortable moment for all involved. Whatever Howard Dean tries to tell you, guns are a conservative thing.[1]

When I was growing up, my family never owned a gun. Since they were immigrants from Norway, it probably never struck my parents as being an option. It would be like owning a wolverine—what's the point in having something in your house that could, and eventually would, kill you? Since we lived in a safe suburb, security never amounted to much more than locking the door when we remembered. While that amounts to, well, liberalism, I guess, there was a strong current of libertarianism in my brother and me. We liked fireworks, particularly the ones that were extraordinarily loud and destructive and purchased at nearby Indian reservations, where there were fewer oppressive governmental regulations. Summertime was filled with our arsenal of firecrackers, bottle rockets, and M-80s raining down upon plastic army men, industrious insects, and each other.

Despite our parents' gun aversion, my brother and I were cheerfully issued BB guns. Rick, six years older than me, got his long before I got mine, but the precedent of him getting one made it an easy argument for me to get one too. Nothing bad ever happened to me in my relationship with weapons and explosives.

Oh, except for the time in fourth grade when I was nearly killed. My friend Dean was staying over one weekend, and looking for something to do, we found my brother's BB gun. Because we were young and dumb and thought the gun empty, we took turns firing blasts of air at each other "to find out how it feels."

We traded shots back and forth, pumping the gun up more and more with each exchange. "Wow, it almost feels like we're being shot!" we said with the kind of giddy stupidity only pre-pubescent boys can muster. Finally Dean pumped the gun seven times and, with my full consent, pointed it directly at my chest and pulled the trigger. Instead of the expected puff of air, I felt a sharp stinging pain and looked down to find a cloud of blood rapidly forming on my white T-shirt. There had been one BB lodged inside the gun that had somehow come loose. As Dean had a complete freak-out meltdown, convinced he had killed his best friend, I watched a trickle of blood roll down my chest and wondered if he might be right about that one. My mom, a registered nurse, sprang into action and, with a mixture of calm, Scandinavian medical demeanor, and Richard Pettyesque driving, rushed me to the hospital. An X-ray revealed the little copper ball to be lodged in some muscle tissue, where it remains to this day. An inch to the left and I'd have been dead at nine years old.

But that was also over twenty-five years ago and the wound has long since healed. By the time of The Experiment, I didn't have much of a strong opinion about gun ownership one way or another. Like abortion and the death penalty, it was an emotionally charged issue that never came up in my life. So during The Experiment, I tried to come at it in my mind from the conservative side. If the American people can be trusted to run their own government, shouldn't they be trusted to own a gun and take on the responsibility of not shooting one another? Does the government have a right to interfere with our choice to own firearms? The Second Amendment is a complete mess in terms of modern applicability,[2] but the idea of gun ownership without a parental government role is a pretty simple conservative notion and one that I could grasp. I certainly didn't share Charlton Heston's "from my cold, dead hands" level of gun enthusiasm,[3] but I wanted a road test.

It can be challenging to find somewhere to shoot a gun, especially if you live in the middle of a city like Seattle and want to do so legally. I found that most Seattle-area ranges were open only to private members and located way out in the woods. After much calling around, I found Wade's Eastside Guns & Indoor Range, located east of Seattle. A fellow named Max answered the phone and I explained that I was looking to fall in love with shooting guns but I had almost no actual experience doing so. In truth, I couldn't say I had *never* shot a gun. When I was twelve years old and on a trip to Colfax, Washington, outside Spokane, with my friend Dean (yep, same one who nearly killed me) and his family, his stepfather taught us how to shoot at clay pigeons. I remember wedging the huge shotgun against my shoulder, feeling my tiny body absorb the massive kickback, and even hitting a few of the flying, yet mercifully inanimate, targets. I think I liked it. Possibly a lot.

I asked Max over the phone if they rented or loaned out guns to people who didn't have their own. He said sure they did, all I needed to do was bring in one of my own guns in order to demonstrate that I was the kind of honest and trustworthy person who had been granted the right to own a gun (or the ability to procure a stolen one, I mused). That might be a problem, I said, since I didn't own a gun. He asked if I didn't have at least *some* sort of gun lying around the house. No, I reiterated, I don't own any guns. But I would very much like to give them a try.

Max, still baffled and now a little annoyed, said there was one other way to do it. I could bring in someone else with me to vouch for my rectitude and provide assurance that I was not some wacko wandering in from the street. Apparently if I was a wacko who wandered in from the street but somehow managed to temporarily befriend another wacko or at least convince another wacko to go with me to the gun range, then everything would be fine and I would be issued firearms. Really, a perfect system.

I called my neighbor Larry. He would be a good prospect, I figured, because he was a young father like me, so our experiences could parallel in some way and he worked from home while his wife, Beth, worked nights so she'd be able to watch their daughter and he'd be available on a moment's notice to go shoot guns. Plus, we'd shot off some fireworks the previous summer together, so it was perfect.[4] Larry was delighted to go shoot guns. If you were working in your home office with your wife and two-year-old running around the house and someone offered you a chance to go shoot guns, wouldn't you say yes?

For wardrobe, I selected Hat Shirt and Rustlers. Larry and I arrived at Wade's an hour later, where Max, beefy, confident, and armed, greeted us. There was a collegial feel at Wade's between customers and staff. It had the same easygoing bonhomie as a well-run Elks lodge, although it lacked both luxury and ornamentation. You weren't there to get comfortable, you were there to practice shooting things/animals/intruders to death. I explained that Larry was here to vouch for my sanity. Larry was a big fan of the historically inept Milwaukee Brewers and wore a baseball cap to that effect to the gun range. Nonetheless, he was accepted as sane and responsible. Max had us sign some paperwork explaining the rules of the gun range:

- Don't wave the gun around when you're trying to illustrate a point.
- Leave the gun on the table when you're not shooting it.
- Don't fire at things in other people's lanes.
- Sign liability waivers that stipulated that just because they were giving us loaded guns, that didn't mean the staff of the range was responsible for what we did with them.

As always in these sorts of situations, you wonder what horrifying incidents took place to lead to such common-sense conditions being included in the agreement.

For a flat fee, Larry and I purchased the right to shoot any gun in the display case along with, presumably, the responsibility not to kill each other. After we explained our naïveté in these matters, Max recommended we begin with a .22-caliber pistol. A nice light gun with minimal kickback that would be easy to handle for us, you know, pussies. I resented the faint but unmistakable patronizing attitude, but then again it was hard to dispute and I thought it was unwise to engage in any kind of argument, given the current surroundings. Max showed us how the gun worked, where the bullets went, how to hold the thing, the right way to aim, and how to squeeze the trigger instead of pulling it. He also recommended that when we shoot the gun, it's probably a better idea not to hit the pipes that run across the length of the shooting lanes or the metal clamps that hold the targets. Doing so, it turns out, would lead to a ricochet effect, with bullets potentially bouncing around the entire facility in unpredictable places. And that could kill us. "Okay," I said, "good to know."

All this explanation was done in the gun shop itself, across (hopefully) thick glass from the shooting range. I figured this would be the introductory session to a more intensive training course that I would probably have to spring extra cash for. But instead, Max zipped the gun up in a little nylon bag, gave us a box of a hundred bullets, and sent us on our way. No certification program, no training, not even any real apparent emphasis on safety. "Libertarian!" I whispered to Larry.

If you've never done it before, there's something ominous about loading a gun. I don't know what it is. Oh, wait, sure I do, it's a deadly weapon that can kill you. I am not the first to make that association at Wade's. Every so often, someone rents a gun, goes out to the range, and kills himself. This was another advantage of being there with Larry. He's a happy guy. Or so I told myself.

We figured out how to load the target onto the clamp, pushed a button to move it into place about twenty yards down the lane. "So how did Max say to load the bullets?" I asked Larry.

"I kind of thought you were paying attention to that part. This was all your idea," he pointed out.

"Yeah. Okay. Well, I think you pull this thing back. And okay, here are the holes that the bullets go into. Those are called the chambers!" My gun knowledge was mostly gleaned from Wu-Tang Clan records.

"Are you sure you have the bullets pointing the right way?" Larry asked.

Once the gun was, we hoped, properly loaded, I was first up. I aimed at the paper target, winced, possibly whimpered a little, and squeezed the trigger. Due to the earmuff headphones, the bang sounded like a dull pop, and as promised, the weapon had minimal kickback. I had more bullets, so I shot some more. Then I kept shooting. I never thought about the instances of accidental death that occur every year or the Freudian implications of men's fascinations with pistols. Upon reflection, the Freudian part was probably a large part of the fun.

Larry took a turn and fired off several rounds. We replaced the targets and each did some more shooting, praising each other for our apparent natural talent. "I'm gonna go get some of those targets that look like people!" Larry shouted, running back to the lobby. We were shouting everything because of our heavy ear protectors. We took turns shooting at the silhouette target where you get more points depending on how much damage would be done to the person being shot, were it a person. So more points for the heart, fewer points for the arms. I've seen enough detective shows to know that the guy shot in the arm always winces but then runs away.

"Get off my lawn!" shouted Larry, experimenting with the whole armed homeowner archetype, "Don't steal my microwave!"

"Dude, good job defending your family," I told him.

"Die! Die!" he screamed, only half in jest, to the inanimate object he was shooting.

"Let's get a bigger gun!" I hollered. We told Max we felt ready for a more aggressive handgun. He went to his collection and brought out a nine-millimeter, presenting it to us like a sommelier handling a bottle of hundred-year-old wine.

We got some more silhouette targets and hit the range again. Again we had a hard time loading the bullets into the darn thing. Unlike most things that aren't working like they should, say a vacuum cleaner, you can't look at a gun from all angles when something is wrong. You always have to hold it away from you so that if it goes off, you don't die. Even a lawn mower is safe once you turn it off. And if something is going wrong with a gun, chances increase that it's going to go off in a way that you really, really don't want it to. A gun is the only thing in the world that can kill you when you're cleaning it. Except bear traps. And Nick Nolte.

The struggle with loading the weapon did not dampen our enthusiasm. Once Larry and I had the puzzle solved (we had not been forceful enough shoving the bullets in), we let the paper intruder have it with our new awesome arsenal of firepower. The nine-millimeter had a more powerful kick and the added awesome feature of making the bullet casings fly up out of the gun into the air when a shot was fired. It looked much more like what shooting a gun by all rights ought to be.

Finally, after a couple hours of this, we retrieved our targets and a few fistfuls of empty bullet casings for souvenirs, zipped the gun into its pouch, thanked Max for his help and guidance, such as it was, and headed home. At this point we were not the same men who had gone into that gun range. We were amped up, like sixteen-year-old snowboarders on our fifth Red Bulls. It was as if each bullet fired had delivered an extra dose of testosterone directly into our bloodstreams.

"Wow! That was really something!" I shouted in Larry's car.

"Yeah! You were really good! At the shooting!" he yelled back.

"Thanks! You too! Did you ever get the feeling that something was going to go wrong and we were going to accidentally kill ourselves or each other?!"

"Yeah! Absolutely! Wasn't that crazy?!"

"Yeah! Ha ha! YEAH!"

"WHOO!"

"Why are we shouting?!"

"I DON'T KNOW!"

We adjourned to our neighborhood pub to have a quick beer before going home to our wives and families. We soon realized that the beer was not merely a social beverage and instead we were using it as a sedative to ease our firearm mania. Larry and I ended up having three pints apiece, way past the accepted urban liberal dad standard, and getting home much later than we originally estimated we would. Jill was upset, having been with the kids all day, knowing only that I had gone to shoot guns, and then finding that I was several hours late and, essentially, missing. Somehow, "Larry and I needed a few beers after the gun range" did not come off like the airtight explanation I had hoped it would be.

Looking back on the issue and on the day's events, I kept returning to the absence of training and preparation I had been given before being turned loose on the shooting range with a live gun. It shocked me that Max hadn't vetted me. Then again, society doesn't either. When you buy a gun, you assume the responsibility of handling it in the right way. Just like at Wade's, they zip it up in a bag, tell you "good luck," and send you out into the world. Most people do the right thing, like I did. Others kill. It was hard to see the gun thing as black-and-white, but having gone out and taken that responsibility, if only for a couple of hours, I felt the appeal of having a private citizen take ownership for the way in which society would be conducted as opposed to having the government come around and decide everything for everyone.

Yep, shooting guns can be kind of a blast, I thought. But I'm not sure I ever need to do it again. I have no problem with other people being gun enthusiasts as long as they're careful, but it's not for me. Especially with the possibility of death lodged in my mind, like a BB surrounded by muscle tissue. The only way it will ever come down to prying a gun from my cold, dead hands is if I ever go back to Wade's and something goes horrifically wrong.

---

1. e.g., Dukakis. Tank.

2. Unless you need to form a militia, in which case it's quite handy indeed.

3. Though I've frequently wondered if anyone actually will do that once Heston dies.

4. I could have called Dean, who still lives in the area, but he would surely have finally killed me once and for all.

# 11

# It **Must** Be Safe
# If It Towers **Above**
# **Everything** Else
## on the Planet

☞

**In which the author, a longtime owner of small and vaguely emasculating economy cars, gets taken out of his comfort zone and into, well, a whole new kind of comfort zone that is, in fact, so incredibly comfortable, it can almost make one, uh, uncomfortable.**

You still get the morning news when you get the news from Fox News Channel, but when you're used to getting it from Steve Inskeep and Renée Montagne on NPR's *Morning Edition,* it's a jarring difference. Urbane, quizzical radio voices replaced by booming well-dressed confident TV people who eerily resembled fraternity guys I knew and feared in college. I tuned in early one morning and learned that Associate Supreme Court Justice Sandra Day O'Connor had announced her resignation.

Within moments of the announcement, conversation on Fox News turned to speculation about O'Connor's replacement, and with that topic, the ghost of Robert Bork entered the room. Bork's 1987 nomination to the court was fought by Democrats who were bothered by, depending on who you ask,

Bork's staunch defense of the Constitution, his ethically questionable firing of Watergate prosecutor Archibald Cox under
orders from Richard Nixon, for whom he was solicitor general,
or his freaky frizzed-out hairdo-and-beard combo. Senator
Ted Kennedy famously declared back then that "Robert Bork's
America is a land in which women would be forced into back-
alley abortions, blacks would sit at segregated lunch counters,
rogue police could break down citizens' doors in midnight
raids, children could not be taught about evolution." I was
nineteen years old at the time and at the apex of my kneejerk
oversimplified worldview, but even still, I thought to myself,
"Oh come on, dude."

We'll never know whether he would have been that bad, of
course, because he was rejected by the Senate. Up until that
point, most Supreme Court nomination hearings were pretty
civil, with presidents putting forth nominees with whom they
were philosophically aligned and Congress confirming them
with little fuss. The Bork nomination changed that and even led
to a new verb being introduced: to *bork,* according to the *New
York Times,* means "to destroy a judicial nominee through a
concerted attack on his character, background and philosophy."
As a consolation for being borked, Bork was granted celebrity
status on the right, becoming a senior fellow at the American
Enterprise Institute, an inevitable television talk-show guest
whenever anything related to the Supreme Court comes up, and
the author of the gloriously titled *Slouching Toward Gomorrah,*
which argues, essentially, and try to follow the logic here, that
we're all a bunch of stinking pervs bound for hell.

My efforts to hear only the conservative point of view during The Experiment received significant help from the produc-
ers at Fox News. The lineup of guests assembled to discuss the
O'Connor vacancy represented a vast array of opinion rang-
ing from conservative to also conservative. Bork was there,

of course, along with Utah senator Orrin Hatch and William Kristol, looking much more cheerful than when I'd talked with him a few weeks before. Guests later in the morning included Senator Jeff Sessions, Republican from Alabama, along with C. Boyden Gray, who had been an adviser to the first President Bush. The Democratic perspective was provided by former Clinton administration member Leon Panetta, who joined the discussion by telephone.

O'Connor's resignation was a huge story since it was the first Supreme Court nomination Bush had been able to make but a hard news story to cover since all you can really do is air a gauzy retrospective of the resigning justice's career and then posit that a conservative president seems likely to nominate a conservative judge. That day, Rush Limbaugh, whom I listened to while running some errands, couldn't even be bothered with something as pedestrian as the Supreme Court and was instead stuck on Karl Rove's comments of the previous week about liberals advocating "therapy and understanding" for the 9/11 terrorists. One of Rush's callers argued that liberals see Guantánamo as a prison and see prison as a form of rehabilitation, which, according to liberals, involves therapy and understanding. Therefore, Rove was right. Rush was delighted, gave the guy four of his special "Club Gitmo" T-shirts, which present the idea that Guantánamo Bay is, in fact, akin to a resort community.

What does it take to get the rush of Rush? As a radio guy, I find plenty to admire about the man's skills. I can marvel at the showmanship. I can notice the highly developed blend of bombast combined with a soupçon (liberal word) of self-mockery. The relentless self-promotion comes off as a bit overbearing, but all animals seek ways to perpetuate their own species, and if there's an author or artist whom you've ever heard of, chances are they are skilled self-promoters as well. If there were no relentless self-promotion, we would all live in a world without Paris Hilton, Donald Trump, or either Simpson sister. And who wants to live in a world like that?

The problem is, Rush so frequently says things that would make all but the most die-hard supporters shout "WHA?!" He has implied that Hillary Clinton had a part in the suicide of White House staffer Vince Foster, he's said that the guards at Abu Ghraib were "just blow[ing] some steam off," and he even dissed Philadelphia Eagles quarterback Donovan McNabb by saying McNabb was overrated but that the media wanted black quarterbacks to succeed. This got him fired from a job he had somehow landed as a football analyst for ESPN. I elected to try to appreciate Rush, the successful broadcaster, while simultaneously ingesting a steady feed of Rush, the conservative philosopher. It helped. A little.

Stopping off for gas, I realized that I had not yet fulfilled part of rule #11: beef jerky, so I spent five bucks on a large bag of Oberto brand jerky. It looked like tree bark but smelled like a dead animal that had been left outside for a few weeks. I hadn't tried the stuff in years, actually, since it's expensive, and to be honest, I always associated beef jerky with the suburban/redneck element of my hometown, which was an element that I had spent most of my postsuburban *New Yorker*–subscribing life distancing myself from. Still, on this morning I was hungry, the sack of jerky was on the passenger side, and with Rush blaring on the car radio, I decided to try it. And Oh My Ever-Loving God. It went straight to my bloodstream. I was flying. This was the best thing I had ever tasted! If this was what conservatives ate, I was ready to sign up for the John Birch Society right then and there.

Now goofy on jerky, I was on the prowl, ready to sample more of what the conservative lifestyle had to offer. It was sunny out and seemed like the perfect day for a drive. Being as how my own car, a 1968 VW Beetle, was often teetering on the brink of total systemic failure, it seemed like the perfect day to test-drive a new ride. Not that I could afford one on my obscure public-radio-host salary. But I wanted at least to try out a vehicle favored by wealthy conservative businessmen and rappers alike: the Cadillac Escalade.

Few things crystallize the liberal-conservative gulf quite like an enormous SUV. Perspectives on them break down roughly this way:

### LIBERAL

SUVs suck down gas to a completely obscene degree.

They tear up the road itself, necessitating more public money to fix roads.

They require more resources to construct, further straining the environment.

They can easily crush smaller cars of the type driven by liberals.

### CONSERVATIVE

It costs a bit more to fill up but we own the world anyway, so no biggie.

Besides, when the rapture comes it won't matter.

Look at this SUV, dude; it's fucking awesome.

They can easily crush smaller cars of the type driven by liberals.

Stopping off at home, I changed into a very nicely crafted preppie conservative outfit: blue polo shirt, tan khakis, Top-Siders. A fresh shave and I was headed out to Bellevue, a suburb east of Seattle.

In his book *The Two Americas,* veteran Democratic pollster Stanley Greenberg characterized Seattle's eastern suburbs, a blend of billionaires, millionaires, and coffee-shop employees, as one of the most contested regions on the national political map. Voters there are regarded as generally liberal on social issues, but also, being in a place known as a technological incubator, very much in favor of free enterprise and economic development. At the Cadillac dealership, however, it appeared to be red state all the way. Wealthy-looking men in sharp suits, some working, some shopping, swarmed around enormous gas-guzzling American cars, plotting ways to unleash the enormous metal monsters on the open road, where their they could chug down natural resources and pump carcinogens into the Pacific Northwest air. Stealthily parking my Bug down the road a piece and scraping off the remains of the Kerry/Edwards sticker Jill had applied long ago, I sauntered into the showroom, a grim, somewhat constipated look on my face that I intended to use as the expression of a successful person, maybe a person with ties to lucrative postinvasion Iraqi reconstruction contracts. As I stood near a floor-model Escalade and crafted my character, I was approached by an eager salesman named Roger. He was well dressed, a little pudgy, and, judging by his almost erotic enthusiasm for Escalades, eager to get a sale.

"Let me tell you something, Roger. I'm looking for the best vehicle option for my needs." My voice was about an octave lower than normal and I reminded myself of Ferris Bueller's friend Cameron when he impersonated Ferris's father in order to help them take a day off. Roger quickly enumerated the Escalade's features in language that I could not particularly follow. Fuel-injected something and dual overhead whatever. Not grasping it wasn't a problem, naturally, since I had neither interest in nor ability to buy the enormous car. So I just stood there squinting.

Finally, I interjected, "I tell you what, Roger. I'm going to cut through all this hoo-hah and rigmarole. Here's what I need: a vehicle that will look impressive when I drive to the golf course to meet associates for a round on the links. I need something that I can take on the occasional hunting trip. And I need to be able to drive it to a PTA meeting and make a big impression there as well, if you get what I mean." He said he did. That made one of us.

I continued, "Now *you* tell *me* why the Escalade is that car." I wasn't quite breaking the rules of The Experiment. It was still me but sort of a performance-art version of me.

After rattling off a series of reasons why the Escalade would, in fact, be the perfect car for my various ruling-class needs, Roger told me about the special pricing offer on all GM cars where customers get the standard employee discount. "Price is not a deciding factor for me," I assured him, adding "at this particular juncture" for no particular reason.

We went out to the lot to select our test-drive car, a navy-blue one. I asked Roger what kind of gas mileage the Escalade got. He told me it generally it gets around eighteen miles to the gallon on the highway and thirteen in the city. I said that I lived in the city, oil had recently topped sixty dollars a barrel, gas prices were as high as they've ever been and showed no signs of dropping anytime soon, and I needed to feel good about the purchase I could potentially be making. Again, I went back to imperative sentences. "Tell me why, with gas prices being what they are, this is an acceptable car to be driving," I commanded him. This is something that I imagine wealthy people who benefit greatly from tax cuts doing a lot, ordering people around. I imagined that this is a tactic that make-believe me learned in make-believe business school.

Roger said there were two reasons that I should buy the Escalade in spite of the high gas prices. First, unlike some higher-end vehicles, the Escalade takes normal unleaded gasoline instead of relying on super unleaded. The difference between super unleaded and regular, factored out over the volume of the entire gas tank, amounts to about twenty-five dollars per fill-up, he said. "So it's like the car is giving you money every time you stop at a gas station," I observed.

"That's exactly right!" he exclaimed. The second reason, he said, was safety. "The Escalade is one of the safest cars you could be driving. And it doesn't matter if the car gets five miles per gallon, your safety is that important."

"How safe is it?"

"It's *very* safe."

"Yes, but by what standard? What is the safety rating? What are the criteria?" I pestered him. The Cameron-pretending-to-be-Ferris's-dad character started to fade a bit as I reverted to talk-show-host mode, chasing a recalcitrant interview subject down a rhetorical alley, seeing what I could expose. Roger said that he actually didn't have that safety information right in front of him but would be happy to get it. Later.

"But really, Roger, all the safety and fuel issues aside, it looks really cool, doesn't it?" I inquired conspiratorially. I was goading him, but it was also, conveniently, true. The car looked great. There was a slight pause. Roger told me that I seemed like a very practical man, so he wasn't going to bring up the Escalade's appearance, but yes sir, it was a beautiful car. With that we took to the road.

There's an important difference between seeing an Escalade on the road and driving one. Seeing one, you might think it's an irresponsible and wasteful indulgence. Driving the $60,000 rig, all you can think is how wonderful it is. Miles above the road, peering down at hapless minivans and compacts, I was seated in soft leather luxury with about a zillion options for how I wanted my seat adjusted and warmed. There was a voice-activated phone system (operating somewhat problematically and requiring Roger to holler insistently at the dashboard, as though at an unruly toddler who refuses to adjust the air-conditioning) and satellite radio with a six-CD sound system. And for the kids, Roger told me, a DVD system as well as plenty of room to stretch out and play video games to keep themselves entertained.

Roger asked what line of work I was in and I told him I was doing some political research at the moment and I'm also in the communications field. "I think I've seen you on TV," he said. "You were on the news, yes? Talking about politics?"

"No." I told him, "That wasn't me. I'm more behind-the-scenes."

"That's too bad. You're a good-looking guy." Which could not be more false. He really wanted that sale.

We went down to a nearby marina where people park their yachts and then go into fancy restaurants to eat steak and drink scotch and talk about the yachts that they just parked. Roger told me to let him out and then drive it around myself for a while. I did. And it produced the effect of being left alone with your teenage girlfriend when her dad leaves the room, exactly the effect Roger was hoping for. I didn't shove my tongue down into any part of the Escalade, but I thought about it. It was intoxicating. I knew on an intellectual level that I could not, would not, should not own this beautiful, beautiful car, but I let myself fall under its devil spell. Would it really be so bad? It's comfortable and it probably won't break down nearly as often as the '68 Beetle. In the space of an hour I had gone from test-driving this thing on a goof to legitimately weighing the possibility of purchasing it. I thought I had been suckering Roger but perhaps the tables were turned. Maybe I could go to law school, become a successful attorney, and then come back here and buy this thing. Would they hold it for me? The price made it prohibitive, of course, but then I remembered: rich people get huge tax breaks!

When I returned, I asked Roger about the deal I heard

about where you can write off much of the cost of the SUV. Originally intended as a tax break for farmers using large vehicles for their work, it also extended to any vehicle with a gross weight plus payload of six thousand pounds or more, including many passenger vehicles like the Escalade. The "SUV Loophole" is often cited by conservationists as being one of the most deplorable tax provisions in history since it encourages the purchase of machines that destroy the environment in multiple ways. Roger said that the loophole applies only if the Escalade is being used for business, not as a personal vehicle. This, he said, is a practice that all of his customers follow very carefully and no one ever cheats. Then he smiled in a really neutral way. I couldn't tell whether he was being completely honest, joking about whether people cheated, or subtly hinting to me about how to swing a sweet deal. I looked at him some more. He smiled some more.

After a few more seconds of that awkward standoff, we drove back to the dealership and Roger did the thing car salesmen do in this situation where they leave you alone for a few minutes to convince yourself to buy the car. I tried to do so, employing the Rich Lowry/Jonah Goldberg "Three-Legged Stool of Conservatism" model. Could I come up with some product or service that the world needed so that I could be rewarded under a classic free-market economic model? That would be a stool leg #1 justification. With the SUV tax loophole, the Bush administration was even giving me some help there as long as it was a business expense. Stool leg #2, the traditional-values and mores one, might be covered under the simple fact that lots of people drive SUVs and that it is therefore traditional. As for stool leg #3, the strong-national-defense one, well, I bet the Escalade could kill some people. You can understand how bad SUVs are for the environment and for our nation's dependence on foreign oil, but when you drive one, you become Alphonso Jackson's daughter and you look out for your own best interests.

Finally, I imagined actually coming home and telling Jill that I really wanted to buy this SUV-asaurus and then I imagined the look on her face. No car was worth seeing that face. That broke the spell.

Sadly, this most bitchin' part of The Experiment was over. I said good-bye to Roger. He offered to let me borrow the behemoth for the weekend to help me make a decision. Free. I knew that either that car or my wife would kill me. No thanks, I said. We exchanged phone numbers and awkwardly promised to stay in touch. I walked around behind a sub shop, ducked behind a bank, sidled along the teriyaki joint, and got back in the Beetle when no one was looking. The car rumbled and sputtered a little more than usual. I think it could smell the Escalade on my clothes.

## PATTON (1970)

**Summary:** General George S. Patton's military career is the subject of this epic biographical film starring George C. Scott, which demonstrates both Patton's tactical brilliance and the mercurial temper that would eventually cost him his job. Germans fear him almost to the point of deification, while America's allies find his crude and aggressive approach distasteful.

## CONSERVATIVE MESSAGES

- War is not so much "hell" as "awesome." It's both sporting and a good bit of fun. Like the board game Risk but with the occasional sucking chest wound. "Rommel, you magnificent bastard. I read your book!" Patton shouts gleefully during maneuvers in North Africa.

- Many within the American military leadership think it's a bad idea to go into Palermo, Italy, but Patton does it anyway. When he gets there, American troops are, in fact, greeted as liberators. See? Sometimes it works.

- The Nazis have been oppressing Africa, Italy, France, Norway, and other places. The Americans want to go in there with tanks and planes but then turn those countries back over to the people to run for themselves. And it works!

- Patton slaps a soldier who he thinks is being cowardly. Then everyone gets all mad at Patton and he's made to apologize, but he was only doing what he thought was necessary to win the war.

## ANTICONSERVATIVE MESSAGES

- Japan and Germany were actively attacking the United States and other nations when we decided to go to war against them.
- Patton speaks French and reads poetry. So he's kind of an "elite."

### OVERALL PERSUASIVENESS SCORE: 76

# 12

# I Put

## the Id in

# Idaho

☞

**In which the author travels somewhere that no one imbibes, then imbibes, and arrives at a somewhat contemplative state of mind, which is understandable given the author's geographic location in a place where there really isn't much to do but drink and think.**

I had known my friend Tina since college. We had been quite close in school and remained so in the years that immediately followed, but after a while we began to drift apart. Happens. While I drifted away from theater, she remained active, doing all number of experimental plays and generally living an artistic, bohemian lifestyle. I got day jobs, mortgages, children, and business cards. Still, friends are friends, and when one gets married, you go.

Tina and her fiancé, Dave, were to be married not in a church but in her mother's sprawling garden/yard that was large enough to accommodate a hundred or so guests. The ceremony was long on anecdote, short on liturgy, and conducted by an actress Tina knew who had been designated as clergy over the Internet specifically for the occasion. There was a lengthy description of how the couple met at a dramatic-movement workshop in Hawaii, and a prolonged discussion of what love was and what it meant and how much everyone should love love. Then another actress got up and read a poem about the Goddess meeting the Buddha. Then actors played folk songs on guitars while standing on balconies and then we had to go stand in a different part of the garden while someone else read something else and there was more folk music. It was a lovely ceremony and a perfect reflection of Tina and Dave's relationship and new life together. It was way outside the traditional oppressive patriarchal construct of what society says a marriage ceremony should be.

I was very uncomfortable. Sure, I was happy for them but it all seemed a little too . . . out there for my tastes. Too hippy, too artsy, too, yep, *liberal.* I stood in the back nursing a gin and tonic (they had no Coors), wearing The Suit, and wishing I had a pew to sit on and a nice dependable pastor who would move things along in the manner to which I was accustomed.

In the car on the way home from the wedding, I agonized to Jill over how square and, yep, conservative I felt at Tina's wedding. "Well, keep in mind," she said, "what our wedding was like." Although it was ten years ago, I remembered it perfectly: we wrote our own vows, it was held in a rented chapel up in the San Juan Islands off the coast near the Canadian border, people dressed casually, and the ceremony was conducted by a chorus of ten of our friends who were appointed as ministers through mail order (the Internet not yet widely used). Tina was one of the ministers. I had long, crazy hair. There was folk music. It was an awful lot like the wedding I had just attended where I had felt so uncomfortable.

"Is The Experiment working that well?" I asked Jill with more than a little concern.

"Maybe you're only doing The Experiment as a way of explaining what was happening to you already," she offered.

I didn't sleep well that night, which was too bad because I had a flight the next morning to Boise.

"Dad! We're having a Fourth of July party!" said Charlie with breathless excitement the next morning as I was packing up to leave. "And everyone from the whole block is coming!"

"Yeah, that's what I heard, buddy. That sounds like so much fun. I wish I could be there."

He was crestfallen. "Where are you going?"

"I'm going to Idaho, Charlie. Remember? We talked about that a while ago."

"Why do you have to go to *Idaho*?" Charlie asked, with a tone of scorn for the state despite having never been there or, I think, ever hearing about it.

"Daddy's trying to learn about some things. I'm trying to change my mind."

"But we're going to have *sparklers*!"

"I know. I still have to go to Idaho."

"But we're having a PARTY!"

"Is President George W. Bush invited?" I asked.

"No! No! He hurts animals! Dad, you were just joking right? RIGHT?!"

"Yes, Charlie. Just joking."

"Bush is NOT coming to OUR PARTY!"

Boise was not my final destination, but it was the best place to fly into before renting a car and driving five hours across most of the southern part of the state to Rexburg, Idaho. The plan was to head for Rexburg for a few days to celebrate America's birthday. Rexburg is the seat of Madison County and the only town in the county with motels. In the 2004 election, Madison County had voted for President George W. Bush at a rate of 92 percent, the highest percentage I could find anywhere in the country. There were counties in Kansas that got into the seventies, a few in Utah that had topped out in the mid to upper eighties, but no one came close to Madison County. What was it about the place that made people so loyal to the conservative candidate?

I tuned in the local talk-radio station in the rental car. There was a feature from the conservative Rutherford Institute that talked about the plight of a Michigan teenager who was suspended from school for posting comments on a satanic Web site. When the hearing for the suspension was held, however, the kid was denied the right to have an attorney present. The Rutherford Institute was siding against the district and in favor of the satanic teen's First and Fifth Amendment. It was the perfect conservative showdown: God versus the founding fathers. And the founding fathers won! "Man, that Constitution is powerful," I thought, "if it can defeat God Himself among conservatives who are willing to be in league with Satan in order to defend it." Should we start capitalizing the pronoun when we talk about the Constitution? So everyone knows how important It is?

In fact, Our Lord Constitution was continuing to receive much praise and adulation on conservative talk shows and righty Web sites in the opening days of speculation about who would replace Sandra Day O'Connor on the Supreme Court. Conservatives, at least the loud ones who land talk shows and get visitors to their Web sites, tended to favor the "originalist" viewpoint on the Constitution and demanded the nomination of judges who treat the Document as absolute truth, as opposed to lefties who think it's open to interpretation. Liberals were talked about as if they intended to disregard the Document and instead roll the whole thing up into a tube and snort coke through it. The station gave way to a scratchy religious sermon, so I switched to numerous country stations instead.

It was a bright sunny day outside, temperature in the nineties, and I had neglected to bring a hat to keep the sun off my blond, balding, Norwegian head, so I stopped at ShopKo. If ShopKo's headwear selection is representative of the allegiances of its customer base, then loyalty is split between America and Nike. I went with a rather busy blue ball cap with an American flag on the front, a yellow ribbon on the side (showing either a commitment to the war in Iraq or devotion to Tony Orlando), and the words SUPPORT OUR TROOPS on the back. I'm always a sucker for imperative sentences. Their raw presumptiveness draws me in every time. You're not only saying what you think, you're demanding that whoever happens to read your sticker or T-shirt or hat does what you tell them. It's the same concept liberals employ when they ask passing motorists to Free Tibet.

Although I was demanding it of others, I didn't really know what "Support Our Troops" meant. I still don't. Is it literal? Like send them stuff to eat and friendly cards and letters? Does it mean not to say anything negative about the war that they're involved with because they might somehow know that I'm saying it and get depressed? Does it mean, as some on the left offer, that we should show our support by bringing them home? Maybe, I thought, if I wear the hat, the meaning will literally drift into the back of my head through osmosis. Or maybe someone reading my hat would then begin supporting the troops right there in front of me and I could see how they do it.

Going to Rexburg requires crossing the Snake River, the same river Evel Knievel attempted to jump back in the days when that kind of thing was widely done. Like Evel, I was making the jump, but I was doing it with my brain instead of a rocket bike. It bears mentioning that the Snake River jumps were the ones where Evel would wipe out every time, shattering his body to bits, and sending him down a long road of bankruptcy and drug abuse. When I arrived, I checked into the Super 8 motel. Rexburg is a town of about seventeen thousand and the home of Brigham Young University's Idaho campus, a school that used to be known as Ricks College. When I was growing up, there were a lot of Mormons in our neighborhood and the consensus was that if you were a Mormon and got good grades, you went to BYU. If you were a not-so-bright Mormon, there was always Ricks. Napoleon Dynamite wore a Ricks College T-shirt in the movie.

I drank a big glass of tap water (maybe something in the water accounts for their voting record, I thought) and set out to explore the town. On my way to the car, I passed through the lobby and fell into a conversation with Jay Peterson, a tall, rangy, avuncular man of about sixty who seemed to approach hotel ownership the way an especially friendly host approaches having houseguests. He was just hanging out in the lobby of his hotel, near the TV showing a ball game and next to the enormous rack of brochures for Idaho tourist attractions (there are some). If you own a hotel in Rexburg and all the rooms are tidy, I guess you might as well just hang out and strike up conversations with your guests. According to Jay, yeah, there wasn't a whole lot else to do on a Sunday. Because of the strong presence of the Mormon Church and the fact that this was a small rural town, he told me I wouldn't find many businesses open. The major exception to this being Wal-Mart, located on the edge of town and providing the people of Rexburg with round-the-clock access to groceries, Puffy America Shirts, and guns. Peterson, who described himself as conservative, was no fan of the retail Godzilla. He didn't like all the overseas manufacturing that Wal-Mart uses and he saw the downside to all those low prices.

"I think I'm patriotic," he said, "but I'm concerned, you know. I know we have to be part of the world, but I also think we have to protect jobs here at home." Cheap products, he explained, are inexpensive because they're poorly made. "We bought some vacuums and carpet cleaners there at the Wal-

Mart and they were cheap. We brought them back here and they lasted maybe two weeks. So I don't care for them."

The big Rexburg Fourth of July parade was not until the next morning, and facing the prospect of just sitting in my room watching Fox News until then, I decided to get out and see the town anyway. It was around three in the afternoon by this point, and as warned, I found almost no open businesses and almost no cars out driving around. It was eerie, like the opening scene in *Vanilla Sky,* where Tom Cruise wanders around a Manhattan that has been deserted. Except this was like *Vanilla Sky* set in Mayberry. Some businesses in Rexburg appeared to be merely closed for the day, others for much longer. A beautiful downtown movie theater, the kind every small town used to have, still advertised *A Series of Unfortunate Events,* even though the movie had closed nine months before. On the other side of the marquee was a real estate sign announcing that the property was available. It was as if the theater owners waited for the most poignant movie title they could find to come along and then fold up shop with that one elegant explanation.

As I drove, I thought about Jay Peterson's comments. Was he a conservative? It's not so much the idea of being opposed to the practices of a Republican donor like Wal-Mart. As Jonah Goldberg beseeched me to, I was not attempting to become a Republican, but Jay Peterson had some stool legs that were battling one another. He was advocating what I suppose were "traditional values" by being in favor of American workers having jobs and American companies selling things made by those American workers. I was still unclear on the values thing, but that seemed both reasonable and traditional. Still, Jay's opinion also seemed to curtail the freedom of businesses to operate without burdensome regulation. He wanted to "protect jobs here at home." If two of Jay's stool legs are twisted around each other, his stool's going to fall over.

I realized that up until this point, I had met very few conservatives who were regular working people. I mean, "Jeff Gannon" was a person and he presumably worked at something or other, but even he would probably agree that there's very little remaining in his life and career that's "regular." And while the philosophy-intensive punditry of Lowry, Kristol, Goldberg, sometimes made some sense, it was all conjecture. They didn't have to watch the trade practices of Wal-Mart impact their towns. They didn't have to follow the ups and downs of manufacturing jobs in order to determine if they would have a job the next day. They could talk about the war but probably didn't know many people fighting in it. Same with me in Seattle, where a job in media and a lot of time spent reading about politics meant that whatever I understood about conservatism, or liberalism for that matter, was not generally based in the real world. It was intellectualized.[1]

I had plenty of time to think about these things while driving around because there was no traffic to navigate, pedestrians to avoid, or anything happening anywhere to anyone. I was trying to get the feeling of the town, trying to figure out what about it made everyone into such President George W. Bush loyalists. I'm pretty sure that even in Bush's own family, he never got 92 percent of the vote.

Then I noticed one open business: Miller's Hideaway Tavern. It was an out-of-the-way place, thus the name, that looked disconcertingly like the bar in the Patrick Swayze movie *Roadhouse,* where Swayze, a brooding philosophical type[2] kicked a lot of people in the face. It was a "get tough on crime" kind of message film.

I had changed into Puffy America Shirt, matched it with the dark blue Rustler jeans, and accessorized with my new American flag/yellow ribbon/USA/Support Our Troops ball cap, henceforth called Flag Hat. I can't say I loved the look, but by now I didn't hate it either. It just felt like my clothes.

Still, I was nervous as I pulled into the Miller's parking lot. As the only bar in town, surely it had concentrated all the rowdy troublemakers in one dangerous place, right? It must have been a place where the pent-up angst of a conservative religious town exploded into an orgy of chair tosses, sex on pool tables, and stabbings, right? Not so much.

There were maybe ten people in the whole joint and everyone seemed to know one another. One guy had a belt buckle that was the size and roughly the shape of a salad plate. I ordered a Coors Light (finding the righty brands was not a problem since the only labels available were Coors, Bud, and their respective lights). The crowd and the town were small enough that the bartender and owner, Jim, only had to be a bartender for a fraction of the time. The rest of the time he was free to drink beer himself. After a while, Jim bought a round for the house, all eleven of us, even me, the stranger at the end of the bar. I got up to use the restroom and realized something horrible: I had walked into the only bar in Rexburg wearing Top-Siders. Patriotic baseball cap, shirt with an eagle flying inside the word *America,* blue jeans, and Top-Siders. I shuffled back to my stool, tucked my feet under it, and kept them safely hidden. I would rather test my bladder than walk around the bar looking like that.

I got to talking to Jim about whether business was booming since he had a booze monopoly on the whole town. "It's actually on the market right now," he said. He thought he might have a buyer ready to go who wanted to knock the thing down and build a strip mall, which is fine with him because he figures, essentially, screw this town. Apparently as Rexburg has grown, the influence of the Mormon Church has grown with it. Ricks College became BYU-Idaho and that meant expansion from a two-year institution to a four-year. That, in turn, has meant a lot more students and a lot more faculty. Great news if you're a Mormon academic, less great if you run a bar.

Jim, a former Mormon himself, said that as the church has expanded and grown in power, the notion of law and order has expanded and grown along with it. On any given day, this town of seventeen thousand is overseen by town police, county cops, Idaho State Patrol, and, in case some form of crime still manages to get past those three layers of law enforcement, the BYU-Idaho security forces are there to swoop in. Jim says what this means for him is a concerted effort to drive his business right out of existence. Cops park outside the bar and pull over anyone coming out of it for suspicion of drunk driving. Almost every time they do, the person is not arrested because they're not actually drunk, but the knowledge that you might get hassled by the cops even when you've done nothing to warrant being stopped has driven Jim's customers away. I wasn't sure how this fit in with conservatism: law and order is a traditional value, but unreasonable search and seizure is prohibited right there in the Constitution in a manner clear enough for an originalist to love. With two competing schools of thought (social conservatives aggressively promoting temperance vs. free-trade capitalism), I wasn't sure whom to try to align myself with. But Jim had bought me a Coors Light, so I listened to him sympathetically. You know how it is when someone finally quits a job they've always hated and they're kind of happy and relieved but still sort of angry that they had spent all that time there? That's what Jim was like with owning a bar in Rexburg.

On top of that loss of revenue from the scared-off customers, Jim said he was feeling the sting of people from the Southwest who used to vacation in Rexburg to escape the summer heat but no longer did because there's nowhere to stay. These were the snowbirds who would come to Rexburg and rent small apartments then be free to spend a warm, sunny, and cheap summer playing golf, stopping in at Miller's, and, in so doing, spend loads of that sweet old-people money. Since

BYU-Idaho went to four years, there are more students sticking around all year long. Since they all need places to live, the school bought up all the buildings and the snowbirds were forced to nest elsewhere.

After the second beer, it was time to leave Miller's and hope that the predatory cops he talked about were not parked outside. By the time this book is published, the bar will almost certainly have been leveled and something, either a strip mall or apartments designated for collegiate Mormons, stands in its place. The remaining drinkers in Rexburg will have nowhere to go to be together.

In total, I had five glasses of Rexburg water that day. Three in the hotel room and two more over at Miller's. Felt no noticeable change as a result. Except maybe one: I was able to walk up to anybody in Rexburg and engage them in conversation. The hotel staff, the bar patrons, the customers and waitresses at the little diner where I had dinner. I would generally initiate the conversations but only when I noticed the Rexburgers poised expectantly nearby, slight smiles on their faces, ready to claim me as their newest friend. They weren't just being nice in that way you're supposed to be with people who are paying you money to do things, they were genuine, willing to get into lengthy conversations with me, a fella they had never met. One guy told me at agonizing length all about the massive Rexburg flood of 1976 when the Teton dam broke and flooded the town with 80 billion gallons of water. He had a friend who was impaled on a tree and gladly told me, a stranger from Seattle with an American-flag cap, all about it. It was an openness and friendliness that you don't find in cities like Seattle or Los Angeles or any other major metropolitan area that I've been to. As I pondered that, I couldn't help notice that all those places tend to be overwhelmingly liberal. Urban sophisticated liberal places: icy jerks. Rural working-class places that vote 92 percent for President George W. Bush: nice folks. Where is the cause and where is the symptom? I wondered. Between conservative voting patterns and magnanimousness, which is the chicken and which is the egg?

Over the man's story of his pal's tree impaling, I heard on the diner's radio the faint strains of Lee Greenwood's inevitable "Proud to Be an American," but later at the same diner, I heard Elvis Costello's "(What's So Funny 'bout) Peace, Love & Understanding?" Although he's not on the approved music list,[3] it had been the second time that day I had heard Elvis Costello. In Idaho. When I was buying Flag Hat at ShopKo in Boise, they were playing "Pump It Up" over the Muzak system. And the previous week, when Larry and I went to the gun range, they were playing, "Alison." Not sure if his line in that song about his aim being true was intentional in the gun-range context. At least they weren't playing "Accidents Will Happen." I was wondering what was up with Elvis Costello and conservative America. Did he get an exemption? Or was this some sort of covert action by a clandestine liberal group that had designs on kidnapping me for a deprogramming?

With so little to do in Rexburg on a Sunday aside from copping free drinks and discussing impalings, it had been a meditative day. My idle thoughts, as they were doing more and more, were starting to make me a little nuts. Fortunately, the next day would provide a whole lot more external stimulation.

---

1. He said, writing a book about it.

2. Yes, in the 1980s, movies were made with Patrick Swayze as a brooding philosophical type, and no, it was not meant ironically.

3. Could someone who wrote a song with that title be anything but a liberal?

## STAND AND DELIVER (1988)

**Summary:** Jaime Escalante takes a job as a math teacher in a tough, inner-city school. His class is made up of gang members, burnouts, and kids who have little history of academic success and few prospects for future accomplishments. In short, they are children left behind. Escalante believes that no child ought to be left behind, and through his passion for learning and some unusual teaching methods, he turns them into the top calculus students in the nation.

## CONSERVATIVE MESSAGES

- Schools don't need more money. They need Edward James Olmos in a comb-over getting all excited and doing some hammy acting.
- If you face down people who oppose you, in this case Lou Diamond Phillips as a gang member who wears a hairnet, they will eventually relent and not shoot you.

## ANTICONSERVATIVE MESSAGES

- The gang members might not constitute a realistic representation of the gang problem, thereby mitigating the extent to which broader lessons about conservatism can be gleaned. In the movie, they mostly look like keyboardists who didn't get invited to Duran Duran tribute band callbacks. Some of them even look like A Flock of Seagulls, which could actually mean that the gang problem is much worse than we ever knew.

- Teachers who can turn inner-city kids into the top students in the world while working for an inner-city teacher's salary are somewhat rare. So as a model for creating an inexpensive successful education model, it might not work. You could take a group of teachers, shave their heads, put them in frumpy outfits, teach them Mexican accents, and shove a fistful of amphetamines down their throats. They would then resemble Edward James Olmos in this movie. But the results would not be the same.

**OVERALL PERSUASIVENESS SCORE: 52**

# 13

# Independence Day

☞

**In which the author experiences the Fourth of July in an environment that is highly familiar in its innate Americana feel, but is certainly alien in the life of a latte-sipping urbanite who tends not to be part of that whole Americana thing.**

Before fully gaining consciousness on the morning of Independence Day, I switched on Fox News. President George W. Bush was giving a speech in Morgantown, West Virginia. He emphasized that even though things might be tough in Iraq what with the seemingly never-ending series of insurgent attacks against American troops and Iraqi civilians and officials, the right thing to do is stand firm.

Lying in bed, I thought about the idea of standing firm. It's not what most people do when things are going poorly. Most people would try to do things in kind of a different way to see if that makes things better. Doing something like that in a situation like the war in Iraq, however, would require coming up with an entirely new plan and admitting that your old plan was not working. Do you pack up and leave the place to collapse into a deadly civil war? Pour in more troops and hope that doesn't lead to an incremental increase in attacks? And where do you get more troops?

It's complex. At least that's how I was used to thinking

about it. But not to President George W. Bush. He says stand firm. Liberal solutions to problems almost always involve complexity. Liberals see a situation like Iraq and say, "Okay, Saddam is brutal and possibly dangerous, but we don't know how dangerous, so let's rely on an international coalition and perhaps look for ways of reconfiguring United Nations pressure to force more compliance with inspectors and also let's gain some intelligence on potential nuclear weapon issues." Conservatives say, "Kill him." Once the war turned into a big mess, liberals revisited all the erroneous reasons for going to war in the first place, condemned the lack of preparation by President George W. Bush, and agonized over potential future scenarios like I did in the previous paragraph. Conservatives said, "Stand firm." Liberals can talk all day long about the web of sociological factors and historical precedents that compose the liberal viewpoint. Conservatives talk about three-legged stools. Not saying one side is right or wrong, but one side is more complex and the other is simpler. The simpler one wins more elections.

There were criticisms I would have made of President George W. Bush's speech, but I decided to stop, to simply listen while lying half asleep there in the motel bed. He talked about all the ways that the terrorists have failed. How they tried to block elections and failed, they tried to prevent the Iraqi governmental council from taking command of the country and they failed. A voice in my head pointed out that there are a lot of deadly and destructive ways that insurgents in Iraq have actually succeeded enormously. Once I blocked that out by using every neuron in my brain to zero right in on the image of President George W. Bush, it was a pleasant and even inspiring speech.

Next up, Fox News had a report on a new museum being built that will teach that God created everything and that, contrary to every scrap of scientific information gathered, nothing evolved from anything else. They interviewed the guy running it, who said at length and with great earnestness that dinosaurs were created on the sixth day, alongside people. Then they had brief quotes from two actual scientists who were only given enough time to say something to the effect of "Huh?!" Then the report was over and we were told to decide for ourselves. It was time to get dressed for the Rexburg Fourth of July parade.

After the previous night's cowboy/Top-Sider debacle, I made sure to be extra careful with dress. "It's not that hard," Jill said with exasperation when I described my gaffe to her over the phone. "You ask yourself, 'Am I white-collar or am I blue-collar?,' then you put the outfit together." She didn't get how hard that was for me since I'm used to just wearing my clothes and I'm never in a position where I must "put the outfit together." So I went back to the previous night's outfit—Puffy America Shirt, Rustlers, Flag Hat, and, making their first appearance in The Experiment, the half-size-too-small cowboy boots.

I figured the parade in a small town would necessarily be a small affair. Nope. Half an hour before the procession was scheduled to begin, the sidewalks were already stacked several rows deep with collapsible lawn chairs, people in flag-themed T-shirts, hats, bandannas, and shorts, and the buzz of excitement over the imminent arrival of their friends and neighbors who would be performing with local gymnastics troupes. Walking up and down Main Street, I quickly realized that there were many more people here at the parade than there were residents of the town. As the county seat, Rexburg had attracted folks from all the nearby hamlets, unincorporated areas, and farms.

The other thing I noticed was the preposterous amount of children in attendance. Rexburg has been sprayed down with a baby hose. Babies being clutched by mothers or dutifully toted by dads, squinting at the bright morning sunlight. Toddlers wandering aimlessly around the small lawn-chair encampments and occasionally wandering into the blocked-off street. Young kids finding friends and happily playing. Preteens and teenagers breaking away from their families to cruise the area on foot with their friends. It seemed like there were many more kids than adults and that every young family had at least three and often many more children. There were even young black kids who appeared to have been adopted by white Madison County families. So not only were they growing in numbers by producing prodigious amounts of offspring, they were importing yet more children from other places. This most President George W. Bush–friendly of all counties in America was amassing an army of like-minded people. Seriously, it's like the Rexburg Flood all over again, except instead of mud, livestock, and debris, it was a torrent of infants that came pouring down Main Street, uprooting trees and buildings, impaling unfortunate campers, and forever altering the town's history.[1]

This is not what you see in cities. In cities, where space is scarce and the cost of living is high, you're more likely to see a young couple with a single baby or toddler encased in thick plastic sheeting in a jogging stroller while Mom or Dad hauls a massive diaper bag overflowing with products to tend to their needs. Some urban couples go way out on a limb and have two kids. In Rexburg, they make massive amounts of babies and let them run free to graze on the open Idaho range. Then they grow up and vote Republican.

While waiting for the parade to start, I walked around downtown and found that even with Wal-Mart's presence it's actually doing better than I was expecting, with plenty of businesses still operating though closed for the day to honor America. Really, Rexburg is pretty much the same as any other small town but differentiates itself in small yet noticeable ways.

A bridal shop is named "Modest Bride," presumably to capture a different share of the market than a shop called "Big Slutty Whore Bride," which does not exist in Rexburg. Down the street from Modest Bride is the Vortex. It claims to be a dance club yet there seems to be no list of upcoming events and as far as I could tell no actual door. Down the other side of the street is an archery shop, which is entirely focused on the bow-hunting aspect of archery. So there are no colorful targets but plenty of fake padded electronic deer. The photography studio has examples of their work in the windows. All the family portraits feature at least four kids.

There was a float dedicated to the various veterans of various wars with many of the veterans riding on it. It looked like a handful of World War II vets, maybe some from Korea, a few from Vietnam, and way in the back a guy who was maybe twenty-two years old, presumably from this most recent war. Unlike his fellow veterans, he was standing up and chucking mints really hard at the crowd and yelling at his friend, "Shut up, Trevor!" Flag Hat, which I was wearing at the time, declared that I supported this fellow, so I guess I wanted Trevor to shut up too. Stupid Trevor!

The National Guard float was a Humvee with a sign that said A TRADITION OF HEROES and then listed WORLD WAR I, WORLD WAR II, KOREA, VIETNAM, and OPERATION IRAQI FREEDOM. There was also a little bit of blank space left over on the float after all the other wars. They've got room to accommodate one more war, but after that they're going to need a new float.

Other parade entries included a salute to the snowbirds, with a small aggregation of elderly people wearing straw hats and waving golf clubs around. There was a float for BYU-Idaho, with students wearing T-shirts emblazoned with the school logo as well as jeans, because apparently despite the sweltering temperatures, the church-dictated dress code would not allow shorts. I saw very few women in actual shorts but plenty of them in capri pants, which extend just below the knee. The capri-pants industry must be very happy with the Mormon Church. I wondered if there was a store in Rexburg that sold only capri pants and nothing else. If this whole conversion thing really took (and I had downed three more glasses of Rexburg tap water that morning), maybe I could move here and open "Johnny M's Capri Pants Emporium" and make a killing.

Something I had not seen since arriving in Rexburg was a Bush/Cheney bumper sticker. True, it was over half a year since the election, but back in Seattle, liberal bastion though it is, I see Bush/Cheney stickers every single day. Defiant, I guess. I also see thousands of Kerry/Edwards stickers left up either out of laziness or resentment. But here it isn't a debate. Nobody here worried too much about political animosity. I guess they were too busy tending to their 756 children.

The parade had been lovely, but I had not had a revelatory conservative moment, a single indicator of what made these people so supportive of President George W. Bush. Perhaps, I hoped, I would find what I needed in the park after the parade. The town had gathered for a big picnic with food booths, high school jazz-band performances, and one of those portable climbing walls that they bring out at fairs that you just know is going to kill someone one of these days. I considered getting some tamales from a possibly immigrant vendor but grabbed a hot dog instead.

I had arranged to meet with Shawn Larsen, the mayor of Rexburg, and I caught up with him at a lemonade stand. The stand was run by the Mayor's Youth Advisory Council, which is trying to raise money to build a new water park since, according to their research, water parks are fun. While William Kristol had looked unnervingly like my late father, Mayor Larsen reminded me of myself. We were roughly the same age, about six feet tall, not skinny but not enormous. Sure, he was a mayor, a Mormon, a lifelong conservative, and he had seven children. But he had been born in Rexburg and I was born in the Seattle area. As Rich Lowry had told me, a lot of political inclination comes from who your family was. Lowry was raised conservative; Kristol's father was one of the founders of neo-conservatism; Jonah Goldberg's dad was a righty writer and his mom convinced Linda Tripp to record her phone calls with Monica Lewinsky. I was brought up by Europeans who were former theater people. The mayor and I sat on a couple of plastic chairs in a shady area in the middle of the park while the town celebration continued all around us.

Larsen grew up in Rexburg before going off to college and, ultimately, a job in Washington, D.C., working in the office of South Carolina senator Strom Thurmond. That's where he was in 2000 when things got ugly in the South Carolina primary between John McCain and George W. Bush. Larsen said the push polling and the controversy that took place there left a bad taste in his mouth, especially when it was between two Republican candidates who shared similar beliefs. He moved back home and was elected mayor in 2003.

Rexburg, he explained, was a family-oriented community, something I had already deduced using my finely honed investigative skills during the parade. It's also an agricultural community. In the fall, school lets out for two weeks to harvest potatoes, an event that Larsen says was highly influential to him. "It taught me the value of work and an understanding of the harvest. Your hard work all summer long pays off in the fall with a good harvest. I guess I relate that back to life. Everything you do takes a lot of work, a lot of effort, but hopefully in the end it all pays off."

Larsen's grandfather was a former mayor of the town and his father was a downtown businessman. He says that the Main

Street area is actually doing better now than it was a few years ago. For a while shoppers were going to the edge of town where the Wal-Mart and a few other chain stores are, but now they're coming back. "I think people like this," he said, gesturing to the crowd. "They like being around other people. They want to see people and visit and have that sense of community. That's what downtowns do in America." Shawn Larsen, Alphonso Jackson, and I had found some common ground.

Unfortunately for Larsen, many of his efforts to improve his town have been undermined by the actions of the administration that voters in the area elected at a rate of 92 percent. Rexburg, and many rural towns like it, have received a lot of funding from Alphonso Jackson and HUD through the Community Development Block Grant program, but massive cuts to that program were proposed in the 2006 federal budget. "I worry that some of Bush's policies negatively affect the city of Rexburg and I don't think that many people in Rexburg understand that," he said, his voice becoming quieter but noticeably more intense.

One of the reasons Rexburg was having such a tough time, he said, was that 92 percent voting rate. The region and, in fact, the entire state of Idaho are so safely Republican that there's never any need for a Republican administration to expend political capital there. No campaign commercials are broadcast, candidates don't show up. It's not like in Florida, where, in the years that followed the disastrous 2000 election, the President George W. Bush administration bent over backward to give the state plenty of attention. I think at one point Dick Cheney gave every Floridian a foot rub. Loyal Idaho, meanwhile, is ignored.

In talking about Idaho's lack of political leverage, Larsen was clearly annoyed. In my efforts to get religion (in the conservative political sense, not the Mormon sense), I had gone to the most Republican county in the nation. In booking some

one-on-one time with the mayor of the county seat, I figured
that it was only logical that I would be talking to the most conservative Republican ever spawned. Had I found a representative of the remaining 8 percent? Had I come all this way to
meet, you know, Rob Reiner?

Not quite. When asked what makes people around there
stick with the Republicans, and in such heavy numbers, he said
that it's because the Democrats are godless and clueless.

"My view [of a good community] is, adults making sure
that they have a concern for children, teaching them to be
honest, to not harm other people, making sure they're connected to the community. It needs to happen with parents and
adults teaching the principles of respect, honesty, and a sense
of community, and caring and respect for other people. I try
to teach my kids to be good citizens. I think the Democratic
Party, they've lost connection with those type of ideals. When
you become the party of unbelievers, or the idea that if you're
a liberal, you're not a believer in a higher power, then you've
lost touch with most Americans. I think most people believe
in God. Then you align yourself with the party that believes in
God." Then you vote Republican.

Having been disillusioned with intraparty nastiness in South
Carolina and dealing with an absence of political clout in his
own state, Larsen was no fan of the Republican Party. He was
glad, he told me, to be mayor in a place where that's a nonpar-
tisan office and he doesn't have to put an *R* or a *D* next to his
name. Like the truck bumpers of his town, he doesn't think
about things in terms of Republican or Democrat or liberal or
conservative. As for the idea that everyone in America is now
binary, liberal OR conservative, Democrat OR Republican, he's
not buying it.

"Yeah, the country's divided, but we're all divided really
close to the center," he said. "I think most people are really

similar on their beliefs about morality, for example. I think you have divisive issues that are divisive and then you have issues that really bring people together. Tom!" He called over an older man who was walking through the park and who had lived in Rexburg his whole life.

After being told that I was there researching the county in light of its voting record, Tom, who looked like what a casting director would hope to get after putting a call out for "grand-fatherly type" or "skinny Wilford Brimley," cited the socially conservative Mormon Church. Not that the church takes a political stand one way or another, but if you're not allowed to wear shorts on a ninety-degree day, chances are you won't be getting behind the party of Ted Kennedy. You'll vote for the guy with the *R* next to his name.

"But I think that's beginning to change," said Tom. I asked him if he was sure about that given that they had voted for President George W. Bush at a rate of 92 percent just a few months before. "I used to be a hundred percent for Bush myself," he said pensively. "Now I'm not so sure. This war . . ." We all stood there in silence for a second, then, with a quick change of the subject to the food at the picnic and a polite farewell, Tom was gone. Man, I thought, when you're the Republican President and you've managed to piss off Rexburg, Idaho, you're doing something wrong.

Many of the soldiers in Iraq come from small towns like this, where their absences, and sometimes their deaths, have a different social impact than soldiers coming from big cities. "One of the sanitation truck drivers is in Iraq right now," said Mayor Larsen, whose manner had begun to get increasingly grave and contemplative, the zillions of flags in the park providing poetic counterpoint as he spoke. Larsen had traveled to Israel and had studied the region extensively. "Personally . . . when is your book coming out?" I told him when, in a gentle tone, sensing that he didn't want to go on the record but his frustration was forcing him to talk anyway.

He chose his words carefully. "I support our troops. I worry about the American lives that are being lost. It seems like the justification for the war has gone from item after item. You don't hear much about Saddam's nuclear capabilities anymore and that was the justification for the war. And I struggle with that."

"Do you think that we went in under less than honest circumstances?" I asked.

"Yes," he replied instantly and fervently.

"Do you think that it was a bad idea to go in?"

"No, I don't think it was a bad idea to go in. That's a tough part of the world. I just don't see it getting any better. Is it our role to see that that part of the world is at peace? I just don't know if it's our role."

So what was next for Larsen? Would we be seeing him in the House? The Senate? What were his ambitions?

"If I run for something else, it probably wouldn't be as a Republican, which in Idaho means that my ambitions will stay ambitions."

"Would you run as a Democrat?" I asked.

He thought to himself in silence for what felt like a week. "Yes," he finally said. He said he feels like the Republicans talk about traditional morality and traditional values on one side of their mouth, then on the other side they cozy up to big business, big contracts, and big money. And the two don't seem to fit really well together.

"Holy crap," I thought, "in my efforts to go conservative, have I turned someone else into a liberal?" But that wasn't it.

Shawn Larsen was a conservative. A pro-government conservative, perhaps, without the small-government stool leg, although he seemed to have two traditional values stool legs to make up for it. He had the strong-military stool leg as well, but he saw it getting sawed off in Iraq. What Shawn Larsen was not was a Republican. His frustration with the war, with the way the President George W. Bush administration was conducting itself, and with the general course of the country had turned him away from the party that was supposed to be the home of conservatives. Less than two weeks earlier, Jonah Goldberg had told me, "I have no great pride in being a Republican. None whatsoever. I have great pride in being a conservative." I think these guys would get along. I liked them both. Mayor Larsen had to get back to the lemonade stand he was helping to run and make small talk with his constituency. I bade him good-bye, we shook hands, and I wandered off through a park full of American flags.

Over at the Madison County Fairgrounds, Fourth of July means rodeo. Since in my own life, nothing ever means rodeo, I decided to check it out. The roads around the fairgrounds were packed with cars by the time I got there a half hour early, and finally I got the clothing right. Yes, it took attending an actual rodeo in order to nail the footwear: cowboy boots (once I had gotten used to the lack of circulation, the boots had become marginally more comfortable). I also went with Flag Hat along with Hat Shirt. If I didn't own a cowboy hat, I could at least draw attention to the subject of hats in general as a means of respecting the concept. A stretch maybe but, dude, I nailed the footwear. I felt good about that.

When I was very young they used to show old reruns of *The Mickey Mouse Club* on TV and I recall some of the sketches/songs/prancings centered on an Old West theme. It was an Old West where everyone was a Mouseketeer and no one swore or shot one another to death with pistols. There were no whores or drunkenness, only happiness and up-tempo cowboy songs.

I didn't know much about the world by that point and even less about the Old West, but even at age seven I knew that this was bullshit in terms of historical accuracy. That same feeling came back at the rodeo in Rexburg. No beer sold at the stadium combined with no one bringing their own boozy accoutrements made for a scene that was rugged and dusty yet still somehow sparkly clean.

I found a seat as close to the middle of the crowd as I could. There were a handful of teenagers on awkward dates that would end with perhaps pecks on cheeks. There were a few older people who had been coming to the rodeo for decades and actually knew a few things about it. Because it was Rexburg, there were also millions of children swarming about. If you found as many termites in your house as there were kids in Rexburg, you'd burn the place down and move away, but kids being cuter than termites, it all created a pleasant small-town feeling. I waved at a few of the people I had met around town, then waved at a few other people who pretended to know me and waved back because that's the polite thing to do.

After the crowning of Little Miss Rodeo (maybe six years old), the pageantry began in earnest with a performance by Rexburg's own Americanas, a group of young women on horseback all decked out in red, white, and blue outfits who had performed at the last presidential inaugural festivities. They performed a precision choreographed horsemanship routine, maneuvering in various patterns while patriotic music played on a tinny PA system that had probably been installed decades ago. Things were going along fine as we watched the horses trudge around in figure eights while someone sang "America the Beautiful."

Then, because it had been a few hours since I had heard it last, Lee Greenwood was back with "Proud to Be an American" playing as part of the routine. After only about a minute, however, the speakers went out. Died in mid-"song." So what did the people of Rexburg do with no Lee Greenwood song? They started singing it themselves, picking up right where the PA went out. Greenwood had cut out after thanking his lucky stars that he lived where he did, his gratitude based on the fact that the flag represents freedom and "they"[2] would be unsuccessful in any attempt to abscond with that symbolic value.

Then, literally without missing a beat, everyone else chimed right in on the chorus, proclaiming, in a kind of breathy sing/chant that they are also proud to be American, remember fallen soldiers, are willing to stand up next to "you"[3] and defend the flag against, uh, someone.

The crowd's memory of the later verses faltered and their voices faded. With the PA system shot, the Americanas and their horses were forced to perform the balance of their routine in solemn silence while a grandstand full of rodeo fans simply stared at them. It was freakin' spooky. The Americanas should make this a regular part of their act.

After the seemingly endless series of rodeo events and a decent fireworks display, the crowd dispersed. I drove the rental car back to the motel and went up to my room. With only an hour left in the Fourth of July, I picked up the pocket-size copy of the Declaration of Independence and Constitution that I obtained at the College Republican Convention and tried to do a good old-fashioned late-night cram session. While everyone on the right rattles on about how great the Constitution is, I found myself becoming enamored of the Declaration. Perhaps it was because I'd had such a long day filled with more ambiguity than you would expect from rural Idaho, but all I could do was be entertained.

"He [King George] has called together legislative bodies at places unusual, uncomfortable, and distant from the depository from their public Records," reads one of the grievances, "for the sole purpose of fatiguing them into compliance with his measures." Translation: he made us show up at these weird places way far away from where we lived so we'd be all be sleepy and go along with whatever he says. I love that that is in the Declaration of Independence simply because of its raw bitchiness. There's also something in the Declaration about "destroying the lives of our people," but it comes after a list of about twenty other complaints. Usually when I want to charge someone with destroying lives, it tends to at least make the top five. "Consanguinity" also comes up. A word that has never been used by anyone since.

I'm not sure if it makes someone more conservative to fall asleep clutching the Declaration of Independence, but that's how I dozed off at the midway point of The Experiment.

---

1. Alternate analogy: Cabbage Patch Kids factory explosion.
2. Who? We're never told.
3. Who is "you"? Wish I could tell you.

# 14

# Rexburg Is

## From Venus,

# Talk Radio

## Is From Mars

☞

**In which the author takes a long drive across the surprisingly extensive southern section of Idaho in the company of Rush Limbaugh and other conservative talk-show hosts. I mean not really with them in the sense of those people being in the car because that would be incredibly difficult to arrange, but just listening to them on the radio.**

Whatever else Rexburg, Idaho, had been in the past few days, it had been pleasant. The cheerful motel owner griping, in the most polite manner possible, about the presence of Wal-Mart in his town and the geopolitical implications of such a store. The proprietor of the only bar in town protesting the dominance of police through gritted teeth and a smile. The conservative-but-not-Republican mayor torpedoing his future political ambitions right into my microphone but doing so in kind of a chipper way while surrounded by the well-scrubbed citizens of his town and their adorable blond toddler army. The rodeo fans drunk on nothing more than rodeo itself. I had begun to grasp the concept of "traditional values" in kind of an abstracted, "know it when I see it" kind of way. Like pornography. But opposite. In my interpretation, it had a lot to do with a level of civility and cooperation that served to enhance the common good.

I dug it. It wasn't like I could feel the traditional values down deep in my soul and they were making me want to have fourteen babies and move to Rexburg (even though you could buy a house there for, like, twelve dollars and I had that sweet capri-pants-store plan), but everyone was nice to me there. Way nicer, as a whole, than people back home in Seattle. Back home, you hold a door open for someone and they walk right on past and don't even say thank you. In Rexburg, they would say thank you, hug you, and possibly give birth to a baby on the spot just for you. I've encountered similar brusqueness in San Francisco, Los Angeles, New York, and Chicago. All voted heavily for Kerry.

My God, I thought in the car leaving town as I pieced this all together, maybe there's something to this. I mean, maybe people are more reserved in cities than in rural communities because there are crooks and con men and traffic in cities, but maybe there's something to the traditional value of being a nice person to your fellow human being. Then I pieced together what other cities or towns I've been in where niceness was a predominant character trait among the general populace. They were all in places like Wellton, Arizona, and Hitchcock, South Dakota. Conservative places. Republican places. I don't like what jerks people are in my hometown, the same town that practically threw a ticker-tape parade when Kucinich came through, months after being eliminated from the presidential race. If people were more pleasant fifty years ago, wouldn't it be nice to return to that part a little? I mean, if we could include blacks and women and The Gay this time around?

Two weeks into it and The Experiment seemed to be working. The abstract social stool leg of traditional values was moving into place. As Rexburg was drawing away from the Republican Party, I was being drawn to Rexburg's way of thinking. I was meeting up with Rexburg at some theoretical place outside party politics and traditional demographics. It was a nice place. If I moved there, my kids would have plenty of people to play with.

Then I turned on Rush Limbaugh. He was railing against all these people who thought that Africa needed some help just because there was all that massive famine and AIDS and malaria and poverty and despair. "If you want to fix Africa, folks, here's how you do it. You get Halliburton to go in there. They'll bring it in on time and under budget." I actually swerved off the road for a second and threw up in my mouth a little. Most of the show was about the upcoming Supreme Court nomination to fill the vacancy left by Sandra Day O'Connor's retirement. For the most part, Rush and other conservative talk-show hosts were upset about the attacks that they expected liberals to make on whomever President George W. Bush nominated to be the next justice. Nobody had actually been nominated yet. Liberals had not attacked anyone yet. Still, the righty hosts were furious about something that had not, in fact, transpired.

This is a tactic that I had begun noticing in conservative talk radio: you take a situation, you imagine the most undesirable thing that a liberal could do in that situation, and then you attack them for doing it even though they haven't. It makes for wonderful drama, packed with familiar characters, conflict, passion, intrigue, and crisis.

The conservative position on Supreme Court vacancies was not about liberalism versus conservatism. It was about "activist" judges versus "originalist" judges. Judges who push their own agenda and then find ways in the Constitution and precedent to back up that opinion are the bad guys. The ones who read the Constitution and then rule based on what's in it are the good guys. So the conservatives aren't saying that all nominees should be conservative. They're saying that all nominees should be originalists. All originalists are, of course, conservatives because the conservative viewpoint states that the Constitution is a not a living document. Judges are there to render decisions based on a strict literalist reading of the document and nothing more or less than that will do. Those old dudes got it precisely right, the thinking goes, and we must not deviate from that lest the republic collapse.

I mock, but in theory it sounds pretty good. To simplify the role of the judiciary to the specific act of following the Constitution as if it were a map is an attractive idea. It removes politics from one of the three branches of government, creates a reliable touchstone from which all wisdom will flow, and ensures that partisanship will not appear in that branch of government. That's an especially attractive idea in light of the fact that the people there are going to be there for a long time, if not until death then at least pretty close to it. There's no electoral referendum on Antonin Scalia or Ruth Bader Ginsburg.

However, the founding fathers did not explicitly anticipate, for instance, privacy issues as they related to the Internet.[1] So when you go to rule on something like that, it's not so easy. Even when they did spell something out, you were stuck with something like the Second Amendment and its lumping in of gun ownership with the maintaining of a state militia. Either they were really lousy writers on that one or they were deliberately fucking with us, condemning us all to argue, while armed, about this thing forever. Verily, we hath been punk'd.

The day's trip across the majority of southern Idaho let me surf across two different streams of Rush, one Laura Ingraham, and even a Dr. Laura. Throughout The Experiment, I generally stuck with the AM dial since switching over to FM could have exposed me to NPR, hip-hop music, and God knows what else. On this particular drive, given my geographical location, none of those things were probably much of a threat, but I didn't risk it.

Sometimes the talk shows weren't expressly bad. Rush riffed on the futility of pop stars at the Live 8 concert and how their own desire for ego gratification and self-promotion (again, this was *Rush* talking about *other people*) made them look ridiculous when they tried to save the world. It was harmless ranting and occasionally kind of funny. Later, Michael Savage bitched about flight delays on his Hawaiian vacation and how he had to take a limo from the resort to the airport and then another limo back to the resort when the flight was postponed but it was okay because the resort gave him his same suite back. Like other hosts, he complains about the "elites" and does not count himself to be among them.

I talked to Jill on the phone from the Boise airport. "I think I'm going to give up the talk radio shows. They're making me less conservative instead of more and blowing all these points I got in Rexburg. They're doing more harm than good!"

"You don't get to pick!" she scolded me. "That's conservatism too! Rush Limbaugh is conservative! You're not in any position to choose what you will or won't listen to. You're not in charge here!"

"But if they're making conservatism less appeal—"

"No. You're not in charge of conservatives. You're trying to let them be in charge of you! More radio."

"Yes, dear."

There was a lot of the world of conservatism that I had become fond of, parts I understood, parts that I admired. I had become a devout reader of *The Weekly Standard* and *The National Review,* where I often found good, interesting reading even if I didn't always buy into the assumptions the arguments were based on. The shooting range was fun. Beef jerky was awesome (and readily available in Idaho, two more large bags consumed in the last three days). Toby Keith and Kid Rock were pretty good. I liked the traditional values of Rexburg (although my leftover belief in women being able to wear shorts on hot days proved unshakable). Not much of a righty dogmatic framework, I grant you, but for two weeks? Not too shabby.

---

1. Keep in mind, they only had the much slower, horse-driven Internet back then; it would be decades before the steam-powered Internet would be introduced.

# 15

## Moral **Turpitude**

### As It **Is** Realized

### Through a Pastime **Not Unlike**

## **Chutes & Ladders**

☞

**In which the author discovers that there are very few
instances in his life of doing What Jesus Would Do and
seeks to rectify that situation, at least kind of, in a sense
anyway, at a Christian book and gift store.**

Back in Seattle, I was ready to resume my normal family life,
at least as normal as it could be after traveling the country without the family, wearing patriotic Wal-Mart wear and
conservative suits, and getting to know the catalog of Kid
Rock on a surprisingly deep level. By the time I woke the next
morning, Charlie and my two-year-old daughter, Kate, were
already up and Jill had gotten breakfast on the table for them
while she tried to convince our surly and recalcitrant children
to get dressed so they could be taken to their respective preschools. The kids, Charlie in particular, had been increasingly
wild when I was around. Whenever Dad wasn't hanging out
with College Republicans or having troubling paternal feelings
about William Kristol, he was flying to Idaho, going to the gun
range, or holed up in the basement watching *Red Dawn*. As a
result, there was more intersibling violence, more inexplicable

tossing of toys and kitchen implements across rooms, more stubbornness in regard to the necessity of actually eating something for breakfast. I knew what this was. This was rebellion. Benign as it seemed now, it would take the form of sex and drugs in the teenage years and must be seen as just as much of a threat. My kids were protesters and I felt about the same way toward them as, say, Dick Cheney feels about, say, protesters. Jill was burned out and frustrated, having been left to run the family on her own. That made the kids scream more and throw more things.

Well, the families of conservatives aren't supposed to look like this. Jam smeared all over kids' faces, pajamas half off, messy hair, flinging spatulas, and punching each other. They're supposed to be well-scrubbed smiling little ambassadors of the future. Like in the closing shot of any Republican politician's campaign ad. I don't know how they pull that off unless they use pharmaceuticals. There's no other way.

It wasn't only our day-to-day routines that had slipped. Our family had been pretty consistent churchgoers, but with the strain of The Experiment, we had stopped going. So in short, my family was crumbling and godless. To attempt to correct this problem was in my best interest because not only was it the right thing to do but because it was also potentially instructive in regard to adhering to traditional values. While Jill took Kate to her preschool, I took Charlie to his. Dressed in the khakis, polo shirt, and Top-Siders, I sort of fit in with the other parents dropping kids off, although the artsy/hippy teachers, who had known me for a while as the casually dressed public-radio dad who drives an old VW Beetle looked quizzical. With a few hours before I had to pick Charlie up, I hit the road, clutching the address of a Christian book and gift store in the area.

I wanted to find a toy, a game, some kind of something that I could use to engage with my own children. The hope was that I could find an item that would be fun for them and also be so jam-packed with traditional values that they would be converted and then maybe I could piggyback on that. Basically, I was trying to turn Bart and Lisa Simpson into Rod and Todd Flanders.

In the car on the way up, Lindsey Graham, the Republican senator with a girly name from South Carolina, was on Hannity talking about the Supreme Court opening. Hannity told Graham that Senators Dianne Feinstein and Joe Biden, both Democrats, had already mentioned the word *filibuster* in discussions about this opening. Graham then gently pointed out that they were saying that only in the context of the phrase *we don't want a filibuster.* I sighed.

Totem Lake is a small suburb northeast of Seattle and northwest of Microsoft's home in Redmond. A strip mall near the freeway is home to a busy Christian gift shop for people who want to buy appropriately nonheathen items. I had to leave Seattle for this mission since as far as I could tell there were no Christian gift shops within the city limits. After some unsuccessful nosing around on my own at the expansive store, I asked Jan the salesperson if she could help me find a good toy for my family that could teach traditional values. My kids, I told her, were four and two years old, and while they were quite bright and could probably handle a board game designed for older players, I really wanted to emphasize the traditional values for them. "Because between you and me," I confided in her, "I'm really not so sure about their morality at this point. So, you know, anything would help."

She showed me a few of the items that might fit the description, including one where you had to build up the "full armor of God." Cute cartoon children were shown on the front wearing heavy breastplates, toting swords, and smiling peacefully (perhaps getting ready to slay cartoon Muslims in a cheerful crusade,

I don't know). The game is based on Ephesians 6:10–20, which tells you to "Stand firm then, with the belt of truth buckled around your waist, with the breastplate of righteousness in place, and with your feet fitted with the readiness that comes from the gospel of peace. In addition to all this, take up the shield of faith, with which you can extinguish all the flaming arrows of the evil one. Take the helmet of salvation and the sword of the Spirit, which is the word of God." So you gather those things up as you play. A little too much metaphor for how young my kids are, I told her, plus Charlie would be demanding the flaming arrows right away. I declined to use the debit card of handiness to purchase this item.

Jan and I considered some Bible Bingo games, but they seemed to teach only about the Bible and not about specific virtues/morals/values. I realize that that's what the Bible is all about, but I didn't want to have to explain the whole gospel to the kids right now, I wanted to get to the moral uprightness right away.[1] We also looked at a Candy Land knockoff based on a cheerful Noah and his happy ark of grinning animals. I asked her why everyone buys Noah toys all the time but no one ever really talks about the rather sad circumstances surrounding the guy. Sure you may have a Noah's Ark mobile dangling above baby's crib, but do you ever explain to baby how God wiped out all the wicked people, leaving them to drown as water rose all around them? And even if they get past that part without shrieking, how do you explain that no one else in the world owned boats? "I guess you could teach about how Noah was good and so God spared him," Jan offered.

"Well yeah, but you still have to explain the drowning deaths of everybody on earth who didn't make it on board. Not ready for it yet," I told her.

"Yeah. I understand."

We finally came to the Fruits of the Spirit board game. I asked Jan what traditional values it taught. She said peace, joy, kindness, that kind of thing. Well, those are good ones, I said. The game is based on a passage in Galatians, which is not an old eighties videogame as I first thought, but rather a whole section in the Bible (considered the Bible of Christianity). It was a sort of a Chutes and Ladders game but without the dizzying thrill of the ladders or the crushing disappointment of the chutes. Just gentle little bridges where the virtuous get to advance a bit farther from time to time. Jan told me that many of the popular Christian board games are based on secular games, but then they pump them full of Jesus. There were Christian versions of Cranium and Trivial Pursuit. There were also instances where the original manufacturer did it themselves with Bible versions of Scattergories and Outburst.

I was almost sold on the Fruits of the Spirit. "Will it make my kids better people?" I asked Jan with much gravity.

"No," she said, "but if you spend time playing a game with your children, *that* will make them better people." I bought it and took it home.

That night after dinner, Charlie and Kate were really eager to play. The way the game works is you get to be one of four beautiful child angels, each depicted on a piece of cardboard held in place by a small plastic clip.[2]

The board shows a path of colored squares winding through areas for Joy, Love, Kindness, Goodness, Faithfulness, Patience, Peace, Gentleness, and, what soon became the Moe family favorite, Self-Control. Each virtue was represented by an icon (heart for love, Bible for faithfulness, and a drippy paintbrush for self-control accompanied by an illustration of an angel resisting the urge to touch wet paint). The game is played by drawing a card, which either guides you to a colored square along the way to the finish line or to the icon of those specific virtues.

My kids saw no reason to wait for a particular card to get them where they wanted to be. "I want the paintbrush! I want Self-Control!" hollered Charlie, who immediately attempted to move his angel card all the way to that square, which is near the finish line. I explained to Charlie what "self-control" actually meant and attempted to explain the whimsical irony of his demand. Since he understood neither irony nor self-control, it didn't register. Meanwhile Kate was fixated on the kitten of kindness (illustration of angel rescuing said kitten from a tree) and was trying to move her angel there. I had to stop her. "KINDNESS!" she screamed, ripping the kitten card from my hand and preparing to punch me with her tiny fist.

"I think we should be able to move our angels to wherever we want them to go!" Charlie hollered. This is often how our games end up in our house, with rules being improvised and players exploring their interests to the detriment of collaborative playtime. Imagine Monopoly done Montessori style.[3]

Not this time. I told Charlie that there are rules in this game. Rules we have to follow if we're going to play the right way. "I'm putting my foot down, son," I told him, sounding like Ward Cleaver and using *son* as a term of direct address for the first time ever. He looked at me. I looked back at him. There was a dramatic pause. He consented to try this "right way" idea of mine.

Once we got started, the game was pretty fun. We drew cards and moved around the board at a nice pace. There were minor glitches. Kate developed the habit of demanding a turn every time, even when she had just had one. I then reminded her of Patience, symbolized by a star on the board. "Remember the star of Patience," I told her. This usually resulted in her demanding to move to that space right away. Charlie meanwhile got greedy with the cards, refusing to return them to the deck once they had been drawn and hording all the cards with icons on them and yanking those ones away from his sister with an angry tug. This would lead to her crying about her lost Kindness or Gentleness card, taken in a way that was neither kind nor gentle. Similar grim fates befell Goodness and Love. But the acrimony between my kids seemed to exist temporarily on a lower frequency than it did most days. That alone was actually a staggering achievement in their social development.

The flaw in the Fruits of the Spirit was that if you drew the right card, you could slide right past all the other virtues and not be bothered with them. Draw Faithfulness or the highly sought-after Self-Control early on and suddenly Patience, Goodness, and Gentleness were someone else's problem. Likewise, you could be close to winning the game and then Love comes along and ruins everything, all your efforts are wasted, and you have to start over. Still, the game gave me a chance to spend time with my kids that didn't involve feeding them, wiping something off a part of their body, or attempting to get an explanation for the outbreak of pro wrestling behavior on each other. Most important, it opened up the dialogue on what traditional values really are. I had a chance to talk to my kids about what peace is, what kindness means, and the apparently hitherto unheard of concept of self-control.

None of the values featured in the game are new notions. They adhere to Judeo-Christian beliefs. In fact, though I'm no theologian, I suspect they adhere to Islamo-Buddhisto-Pagan(o?)-Judeo-Christian beliefs. They're just good ideas that help you better navigate your way through society. I guess that's conservative. But how can anyone not like that?

1. Yeah, I know all sorts of warning bells are going off for Christians right now.
2. No explanation of how these poor kids died. And no explanation demanded. Thankfully.
3. Which would be Marxism.

# 16

# Climbing
# Inside the Radio

In which the author goes into the studio of a
conservative talk radio program where, instead of
angrily debating the host as the author normally might do
in a car, he sits and listens and tries to learn things.

July 7 was just another day to me when I woke up. I showered, got dressed, and headed out the door to run some errands and go to an appointment I had set up. I had no idea that terrorists had set off a series of bombs in the London Underground, killing fifty-six people and injuring hundreds more. I got to hear about them from Rush Limbaugh on the radio when I got in the car. Within about eight words of describing the events, Rush arrived at the word *leftist.* Leftists were partially responsible for the attacks, he indicated, because of lax immigration policies and an overly solicitous approach to radical Muslims. Funny, I thought, I could have sworn it was the assholes with bombs that were responsible. Rush Limbaugh was becoming like the crazy new girlfriend you have that's constantly ripping your friends and is high on Oxycontin. But you stay with her because . . . because . . . I have no idea why you stay with her.[1]

I would have only a limited time with Rush, however, because I had to get ready to spend the day with one of his friends. Michael Medved is the host of the imaginatively named *Michael Medved Show,* produced a couple of miles from my own station in Seattle and syndicated to markets around the country. Not only was he in my town as well as being a colleague (if a major national commercial host and a minor local public-radio host could be colleagues), Medved had a trajectory that I found intriguing. He had attended Yale undergrad at the same time as John Kerry and President George W. Bush, and attended Yale Law School with Hillary Rodham, whom he liked, and Bill Clinton, whom he didn't. A job in the office of a congressman Medved describes as a "Stalinist Democrat" followed, as did a career as a film critic on PBS, CNN, and in the *New York Post* and ultimately a conservative-talk-radio hosting job.

I had written to him, introducing myself, explaining The Experiment, and asked if I could be an intern/gofer on his show for a day in order to learn the inner workings of conservative talk radio. He said that they never had enough stuff to keep their actual intern busy, but I was welcome to come in and sit in the studio with him while he did the show one day.

As I sat in the lobby at the Seattle Entercom studios, its boastful walls festooned with awards that didn't matter much to anyone aside from the people receiving them, I wondered how Medved, whose show I had barely ever heard, would respond to the bombings. Would it be Kerry's fault? Bill Clinton's? Hillary's? Chelsea's? Moveon.org's? In exorcising all things potentially liberal from my life for the purposes of The Experiment, I had also exorcised most news outlets, so I didn't really know anything about what had happened in London outside of what Rush had been telling me. I tried to create my own mental NPR newscast to hear about the bombings. I think I had Corey Flintoff reading it. Sylvia Poggioli reported. It was insightful and in-depth and didn't blame any political philosophy for anything. Yes, my make-believe newscast may have sounded a touch pedantic, but at least it was saying smart things.

Moments before the show went on the air at noon, Jeremy,

the show's engineer, young, casually dressed, goateed, identical to a thousand liberals I knew but different, came to get me. I was ushered into a cramped-air studio where Medved was sifting through piles of notes that were already heavily speckled in green highlighter. For those who didn't watch movie preview shows on PBS in the 1980s, Medved has thick black hair and a mustache and looks like the offspring of an unholy coupling of Mr. Kotter and Horshack. As is the case with most radio shows, Medved's studio is considerably less grand than one might imagine listening at home or in the car. A smallish room dominated by a mixing board, a computer monitor with a list of callers (names, locations, brief summary of the point they told the producer they wanted to make), some chairs, some headphones, a few microphones. You could fit it all in a picnic shelter at any county park. Medved greeted me warmly, shook my hand, and offered me a seat. He made a crack to Jeremy about how I was here to write about them and bring them all down. Not my intention at all, but a flattering comment from a gracious host.

Just as in the studio at my station, there's a TV, but whereas we usually keep ours tuned to an external camera so we can comment on weather or traffic, Medved's was tuned to cable-television news with the sound turned down. Not Fox, CNN, the station variously called the Communist News Network and the Clinton News Network by righty critics. "Wait a second, did I wander into Al Franken's studio by accident?" I mused.

"Another great day in this greatest nation on God's green earth," Medved said when the on-air light went on. "Yes, it is a great day, though not great in the sense of wonderful, great in the sense of significant." Obviously the London attacks would be the topic on Medved's show just as it had been on Rush's and would be on everyone's. Just as obviously, everyone wanted to learn more about what was actually happening with the story instead of merely producing call-in talk shows to discuss it. So there was CNN. During Medved's frequent breaks (the average commercial radio hour has only about thirty-five minutes' worth of actual programming), he and Jeremy would turn up the volume on the TV to learn if there were any new developments. At one point, about midway through the first hour, Medved asked Jeremy to switch to Fox News to see how they were covering the story. "Alas," I thought, "that's the last I'll be seeing of CNN for a few more weeks." After a few seconds they switched it back.

Turns out they both dislike Fox News but were morbidly interested in how poor Fox's coverage was. Before The Experiment, I had believed that conservatives in America were more or less of one mind. Not so. There was dissension, rivalry, and deep division.[2]

During another break, Medved confided to me that he listens to NPR's *Morning Edition* and *All Things Considered* every day. He considers NPR the best source of news available on the radio and better than newspapers in many cases. This was not a shocking thing to hear someone say in Seattle. I happened to think he was right about NPR and I know tons of people who feel the exact same way. But most of those people are liberals or employees of public radio itself and none of them are conservative commercial talk radio hosts. An hour had gone by, and though I found some of his reasoning flawed and I wasn't arriving at the same conclusions he was about the debate over terrorism, I was also having a pleasant time hanging out with Michael Medved, a phrase I am stunned to see my own fingers type.

Medved's only guest that day was Robert Spencer, a professional right-wing talk-show guest, occasional author, and director of an organization called Jihad Watch. Spencer's perspective and Medved's as well was that the attacks were a reminder of the real and ongoing threat of "Islamofascism," a compound word popular among conservatives. Both also believed that this form of militant Islam is far from uncommon and is in fact on the rise everywhere. Also that President George W. Bush and Tony Blair are really terrific.

While his touch was lighter than Limbaugh's, Medved did go after liberals. After attacking some of the more spurious logic from random message-board posters on the lefty group blog Daily Kos, Medved went on the attack against George Galloway, a British MP who has become a hero among some on the left despite being pretty much drummed out of the liberal Labour Party in his own country. It would be putting it extraordinarily mildly to say that Galloway was against the war. He had lambasted Blair and President George W. Bush several times on the subject, had a close relationship with Iraqi officials, and had professed his admiration for Saddam Hussein. In light of the London bombings, Galloway had made a statement condemning the actions but saying that they are a result of the ill will stirred up by American and British military actions. Here's what Michael Medved did with that statement:

"[Galloway] said 'no one can condone acts of violence aimed at working people going about their daily lives.' What does that mean? It's okay to have acts of violence against the kinds of people in the World Trade Center because they're not working people? 'Cause they're bankers? Isn't that outrageous? 'No one can condone acts of violence aimed at working people going about their daily lives'? And then he goes on. 'They have not been a party to, nor have they been responsible for, the decisions of their government. They are entirely innocent. And we condemn those that have killed or injured them.' Meaning if the terrorists had struck at Blair or his government, he would support the terrorists in that? That then violence would be justified? What about murdering Mrs. Blair? Or he has a baby, they have a baby at home, remember? The Blairs have a kid who is I think four years old now? What if the terrorists killed Sheree Blair and the baby? Is that okay? Because they're not entirely innocent?"

It was the old reductio ad absurdum where you take a position for the sake of argument, arrive at a ridiculous result, and then use that ridiculous result to discredit the original position. The same way that conservatives argue that if The Gay is allowed to marry, before long people will be marrying logs and buildings and the 1975 Steelers. It would be ridiculous to marry the Steelers,[3] therefore it's ridiculous for two consenting adults in a long-term stable relationship to decide for themselves that they want to get married and register for a toaster. The reductio ad absurdum, besides sounding like a Harry Potter spell, has the words *reduce* and *absurd* built right into it. Not the most airtight logic in the world, but it makes for great radio.

Still, I wasn't there to judge even though it would be super easy to do. I was there to try to get on board. So in order to zero in on the day's attacks, I summoned up the feeling I had after 9/11 when, like loads of people, I thought, "Let's go get those bastards." Like Toby Keith, I wanted to put a boot in their collective asses.

Instead of being merely an echo chamber of die-hard fans giving the host "megadittoes" as is the case on the Limbaugh program, Medved repeatedly asked for people who disagreed with him to call in and tell him why, particularly because on his show, Thursdays are "Disagreement Days." Perhaps it was the challenge of coming up with a counterargument to "terrorists are bad," but the majority of disagreement calls that came in were from conspiracy theorists and psychotic militants.

Carlos in Carson City, California, called in to argue that Saddam Hussein was not a real enemy to the United States and had in fact been propped up by the U.S. government for years with money and supplies. Medved pointed out that at the time, there was a war going on between Iraq and Iran and the thinking on the part of the Carter and Reagan administrations was that we were better off backing Iraq in that particular conflict. Carlos went on to say that America has always created fake enemies in order to prop up the military-industrial complex. But as he talked, it became apparent that Carlos was taking this pretty familiar liberal argument in a different direction, namely to a bunker wrapped in tinfoil. The fake enemies in Carlos's estimation included Hitler and Stalin, who were concoctions of the secret cabal of Jewish bankers that plotted to rule the world in cahoots with the Illuminati and wealthy families like the Du Ponts and the Rockefellers. It eventually came around to Yale University, thought, by people who think such things, to be the epicenter of this conspiracy. Carlos noted that Medved had gone to Yale, which the host conceded while pointing out that he had never belonged to the secret Skull & Bones society. We were so far through the looking glass at this point that I had almost forgotten Carlos's original and valid point about American funding of world leaders it later opposed.

Carlos was a liberal, I gueeessss, but either I had really gone off the deep end of conservatism and all liberalism sounded like paranoid schizophrenic ranting to me or Medved was airing only whack jobs that speak poorly on behalf of the left in order to make his largely conservative audience feel better about the comparative sanity of their own beliefs.

Tony from Minneapolis provided a different take, lending the conversation some valuable insight on the oft-overlooked notion of Michael Medved's irrevocable future in the underworld. "When you die, you know where you're going, 'cause you're serving the master, which is Satan," he said, helpfully adding, "and I may sound crazy but you know exactly what I'm talking about." Like the rest of the planet, Medved did not know what Tony was talking about. Tony explained that Medved has been rewarded with cars and houses and a radio show because he's been serving Satan the whole time. I had never seen such provisions in public-radio contracts, but maybe perks are different in the commercial sector.

A caller from Cincinnati named Black Lion spoke with the level of subtlety and nuance that you expect from someone named Black Lion. He said that the bombings were not a terrorist attack, but a reasonable response to the invasion of Iraq. "If an oppressor comes in and uses bombs against them and their women and children and their civilians," he said, "male and female alike, I think that it would be best for the same tools to be used against the occupiers.

"This is roosters coming home to roost," Black Lion stated.

His belief was that the London bombings were nothing compared to the crimes of the slave trade. Black Lion said, "Now, for five, six, seven hundred years, little Britain and so-called America has been going from coast to coast, from the jungles of Asia to the shores of Africa to the desert to Iraq, murdering, killing, raping, robbing." He then blamed Medved for slavery and announced his plan to go live in North Korea or China or something. Black Lion was to liberalism what White Lion is to heavy metal.

Where do they get these loons? And where were the reasoned, intelligent lefty voices? Did they get screened out by the intern taking calls? Or did the call never arrive in the first place? You have to be a certain kind of person to listen to conservative talk shows. Then you have to be a subset of that group to be the kind of person who calls in to those talk shows, looking to get your voice heard by a mass audience. Then, after those filtering mechanisms, you have to be a liberal listening to conservative talk radio and choose to call in and match up against someone who makes a living (in league with Satan or not) arguing ideological positions. Not a very good sample group remains at that point, yet it is these spokespeople that form the image of liberals to conservatives who hang on the words of talk-show hosts.

Still, unless Howard Zinn was on line three and not being allowed on the air, was any of this Medved's fault? He's doing his job, hosting a show that supports his genuine beliefs and inviting liberals to call and opine. When they do, he lets them talk, doesn't cut them off, and even asks them follow-up questions in the same manner one would talk to a sane person.

Thus, you arrive at conservative talk radio: limited by commercial constraints to frequent breaks that interrupt attempts at in-depth conversation, favored by callers who agree with the host, and stuck with dissenting opinions by Black Lion.

In the studio's break room after the show, I spoke with Medved about the format. I attempted to ask a question about "conservative talk radio" but he challenged that term. "One of the things that people like to do is talk about 'conservative talk radio and Fox News.' First of all, we're not fans of Fox News on my show. I know a lot of other talk-show hosts who are not fans of Fox News and all of us are different. Talking about conservative talk radio as a unit is a mirror image of people talking about 'the liberal media.' There is a difference between Pacifica [Radio] and NPR, thank God.

"I hope that everybody would make a distinction between *The Michael Medved Show* and *The Michael Savage Show.* Savage is an embarrassment. He's an embarrassment to our industry. He's an embarrassment to the human race. Basically, I don't think he's for real. I really do believe the stuff I say on the air, I care about it.

It's who I am. It's my life. I don't think that's true for him, I think he's playing a role on the radio. I think he's created a character and maybe that character has taken over." As we spoke, Michael Savage's show was on the air, carried by the same station immediately after Medved. Medved pointed to Dennis Prager and Hugh Hewitt as reasonable conservative talk-show hosts who also happen to be distributed by the same syndicate as Medved and he defended Limbaugh, sort of. "Occasionally there will be stuff that I'll hear on the Limbaugh show that will embarrass me, but I think it's wrong to equate Rush and Michael Savage. I think it's wrong to equate Rush and Sean Hannity. Rush is sophisticated, extremely well-read, smart, and don't ask me about that other show."

"You mean Hannity?" I asked.

"It's different," he said, making clear that "different" was not a compliment.

In addition to the benefit of providing him a nice career, Medved and his dark master/supervisor Satan believe that conservative-talk-radio programs provide a valuable service for society in general and not solely for the like-minded. One of the reasons he said he found The Experiment so interesting is that "the liberal world can be totally self-enclosed and self-referential. It's simply isn't possible in the other direction. If you don't tune into talk radio, it's entirely possible to avoid conservative messages in the culture. They really don't exist. What's a conservative TV show?"

I couldn't think of one.[4] I couldn't think of a liberal one either. "What's a liberal TV show?" I asked.

"*Will and Grace. West Wing.* Most everything else."

*Will and Grace?* "Because there's gay people on it?"

"It's the most wholesome and positive possible view of gay life," he said. "We don't have a TV at home, but I was given an episode I watched on tape where they confront a, quote, ex-gay minister. Which is rather savagely ridiculed."

"Oh. How about movies? *Forrest Gump* has been called a conservative movie." I had recently seen this one and related the scenes where the antiwar protesters are really cruel and how the character of Jenny is killed, essentially, by the free-wheeling liberal culture she embraces.

"But Forrest Gump is also mentally challenged," Medved pointed out. Our conversation kind of petered out from there.

A couple of refreshing things had happened, though. First, I made peace with the fact that conservative talk radio is a show. It mostly consists of like-minded hosts, guests, and callers verbally patting one another on the back with the occasional Black Lion thrown in to make the dominant opinion seem eminently reasonable. It's not really a discussion so much as a continual presentation of a perspective. For liberals, it can be an instructive way of hearing that sort of opinion if you can stop shrieking.

Second, just as Jonah Goldberg had granted me the patronage of Starbucks, so had Michael Medved allowed me to be embarrassed whenever hateful bile came shooting out of my radio. And that was a relief.

It also made me wonder about the power of actually talking to people. I had read things written by Rich Lowry, Bill Kristol, and Jonah Goldberg that raised my blood pressure, but when I met them, they were nice guys and possessed of the ability to reasonably present their opinions. The one conservative-talk-radio host I met had turned out to be polite, gracious, hospitable, and a decent fellow. Even "Jeff Gannon" was an okay dude to have a cup of tea with. Is direct contact an efficient means of bringing our nation together or am I a wimp whose convictions melt away as soon as someone shakes my hand? Hard to say. All I knew for sure was that meeting real people and really listening to them made me more of a believer in constructive dialogue and more of an agnostic when it came to the bitter conflict allegedly tearing our country apart.

1. My wife would like me to point out that she has never used Oxycontin and that the above ought not be taken literally.

2. And oh my goodness, righties, don't fret that you're the only ones bickering. The left invented that dance.

3. Even though the strength of that defensive line would be an asset to any marriage.

4. But I sure did later. *Everybody Loves Raymond, Touched by an Angel, Matlock, Happy Days, The Apprentice,* almost every other show with a nuclear family or a cop or a detective, so therefore, almost every show ever put on TV.

## FORREST GUMP (1994)

**Summary:** Forrest is a simple and honest man who accidentally plays a part in many of the major cultural and political events of the twentieth century. He teaches Elvis Presley how to dance, inspires the John Lennon song "Imagine," invents the smiley-face icon, coins the phrase *shit happens,* and meets with several presidents of the United States. Along the way, he pursues his true love, goes to Vietnam, brawls with Black Panthers, becomes a shrimp magnate, and seems more or less happy by the end.

### CONSERVATIVE MESSAGES

- Forrest Gump is not one of the elites that conservatives so often warn against.
- He goes to Vietnam when asked by his country to do so.
- He makes his fortune through capitalism.
- He stands up to radicals and hippies.

### ANTICONSERVATIVE MESSAGES

- He is also pretty much an idiot.

**OVERALL PERSUASIVENESS SCORE: 60**

# 17

# Think Globally, Conservatize Locally

☞

**In which the author explores his own community and seeks to find the often minute distinction between "social experimentalist" and "freaky pariah," all the while trying to earn the admiration of his children or at least trying not to earn further scorn from them.**

In the days since I'd returned from Rexburg, my temper had grown shorter. In my best moments, I found conservatism to be a fascinating blend of surprising ambiguities; in my worst moments, it was something I couldn't even understand, much less be won over by. The isolated moments of irritability had started getting larger, more frequent, and connected to one another. Instead of becoming moments, they were now parts of my personality. The kids would be acting impossible, at the same steady level of impossibleness at which they always acted, and it would drive. Me. Nuts. I would either end up scolding them about something completely stupid and unimportant or going out on the porch to grit my teeth and count to ten. And the incident that triggered it would be, like, some Cheerios getting spilled. Meanwhile, with all the travel and the time spent watching conservative films, I had packed on some extra pounds, which look extra unflattering in Wal-Mart clothes. My sleep was fitful and punctuated once by, and I'm completely serious about this, a dream about Bill O'Reilly and Sean Hannity chasing me around, but instead of being people they were monster trucks. Jill was not avoiding me necessarily, but didn't seem particularly eager to deal with me on a very personal level. Couldn't blame her for that. I didn't want to deal with me either, but I didn't have a choice.

Liberal explanation of why I was becoming a jerk: the toxic anger of conservatism was poisoning my mind, bringing about a raging dementia.

Conservative explanation of why I was becoming a jerk: the truth was finally being revealed to me, but I was being held back from accepting it. Much like the Iraqi insurgents, my mind was not greeting conservatism as a liberator; instead my mind was setting off roadside bombs.

My explanation: I, uh, did not have one. Not a good one. I would think that anyone might be disoriented and possibly testy, though, if that which they had known of their life was in large part taken away. My five days a week at the public-radio station, gone. My habit of gathering news and information from NPR and the *New York Times* and *Washington Post* online, gone. My political discussions with Jill, with friends, with coworkers, all gone. Even my cultural life was no longer with me. I had bought this wonderful record by a Portland band called the Decemberists right before The Experiment began, listened to it obsessively in the days leading up to my flight to D.C., but because it had an antiwar song (and how likely was it, really, that an alternative band from Portland was composed of die-hard supporters of President George W. Bush?), I had to yank it from my playlist/life.

So perhaps all that made me a tad irritable to my wife and kids. Or more than a tad. More like a truckload. In June, I had been Kucinich. By July, I was becoming Dick Cheney, and Jill, my unfortunate Pat Leahy. I had also developed a tic. In times of stress, my right eye started to jerk around and close up, the muscles around that part of my face clenching up with it.

When you're of a certain age, you get into some routines in the morning. Actually even if you're young, you get into a routine. Charlie likes to start the day with a cup of milk. I enjoy coffee and the newspaper. But aside from the Family Research Council–approved sports page, I would not read the Seattle newspapers, and unless I felt like going online for NewsMax or FoxNews.com early in the morning, I was out of luck. So I sat at the table at six o'clock the next morning with a cup of freshly brewed coffee and an Ann Coulter book from the library. Scowling.

"You need to watch the kids today," said my wife, Jill, to whose name I had attached the prefix *long-suffering*. "You need to watch the kids and then later go get some exercise," she said, looking at the imprint of my gut under the "Don't Mess With Texas" T-shirt.

"But . . . but . . . there's an Ann Coulter book that I need to . . . William F. Buckley autobiogra . . . things to do," I protested meekly.

"Your family needs you."

That was that.

Charlie was playing on his Leap Pad, a plastic notepad toy that comes with booklets that help you follow along with stories or play games. It's a training device for both school and video-game dependency. The game he was playing required him to pick a dinosaur to represent himself, Pentaceratops for instance, and then encounter another dinosaur like T. rex and decide what to do. I overheard him select the "run away" option pretty often instead of choosing to stay and fight. Finally, I decided that it was time for Charlie to stay and fight the other dinosaurs, to recognize that they presented an immediate threat, and war was the best option (especially after how T. rex treated the weapons

inspectors). He humored me. Or, worse, trusted me. It turned out
that his dinosaur was not properly equipped for such battles and
would routinely get killed. Later, Charlie showed me how sometimes he elects to run away or "make friends" with the other dinosaurs. Suddenly Coelophysis or Allosaurus or some other predator
would become his dinosaur's pal and no one was getting killed
when they didn't have to.

After breakfast and another quick round of the Fruits of
the Spirit, I took the kids up to the business district of the
neighborhood where I live. Lots of little independent shops,
a cool indie record store that also serves as a café, a toy store
with educational and nonviolent toys, more than a couple of
Thai restaurants. I wore Puffy America Shirt and Flag Hat.
Since Dad was wearing a hat, the kids decided they needed to
as well. Kate went with a floppy Washington Nationals hat that
I had picked up at the game and Charlie chose my NASCAR
cap. Here comes the Moe family.

Outside the seedy dive bar that has somehow survived the
massive tsunami of gentrification, a toothless boozehound ready
to start his day's drinking at eleven-thirty saw Charlie's hat.
"Awright NASCAR!" he said, "Lookin' good, buddy! Who's
yer favorite driver?" My poor urban liberal son had absolutely
no idea what to do with the question. I wanted to help Charlie
out but kind of froze as well. The only driver I could think of
was Jeff Gordon, but I know enough to realize that Gordon
is more hated than loved.[1] I almost said Charlie liked Ricky
Martin, which sounds like the name of a race-car driver but is
actually a Latin pop star of indeterminate sexuality. There was
Dale Earnhardt, "The Intimidator," but he died at Daytona a
while ago and to claim that a dead man was my son's favorite
would be a bit creepy, even for barflies. Finally, I knew what to
do. "You like Junior, don't you Charlie? Dale Junior?" Charlie
looked at me without a droplet of comprehension. It was if

I had said, "You like monkey helicopter, don't you Charlie? Toothpaste watermelon monkey helicopter?"

"I got a Junior T-shirt," the man said. "It's at home, though, right now, I hain't wearing it right now," a clarification the necessity of which he and I disagreed about. I did notice that the boozehound and I were wearing almost the exact same baseball cap, blue with a big flag on the front. Mine had the yellow ribbon, though, meaning that I loved soldiers more.

Next to the bar was our actual destination, a local bakery where everyone knew the kids from their frequent visits with Jill. Inside, I ran into Laura, an old coworker from about four years ago who was in buying bread. Charlie and Kate started running around the place and Laura was on her way out, so we only got to visit for half a minute. I had no opportunity to explain my apparel, leaving her to look at me and wonder if she ever really knew me in the first place and if she had been incorrectly assessing NPR all those years. She seemed pretty eager to get away from me. I really need to e-mail her.

This, I finally realized, was the fulfillment of the wish Jill had made during our original shopping trip, that friends would see me and wonder if I was okay.

I wonder what it was I had always found so objectionable over the years about the blue-collar heart-on-sleeve conservative image created by the discount retail apparel. I don't think that it was the soft nationalism of the clothing's sentiment but instead the garishness of the clothing itself that made me reluctant to wear it: the bright colors, the cheap quality of the garments' construction. When you really look at it—and no disrespect intended toward America here—but have you ever thought about the flag purely in terms of design? I mean, leaving aside any feeling you have about America, the actual flag is kind of garish. Stars? Stripes? Together? Am I being too metrosexual and Seattlesque by even thinking this?

After dropping the kids off at home to spare them further psychological damage, I went to the gym, where on Fox News there was a story about stocks doing well despite the terrorist attacks in London. The text at the bottom of the screen read "Stocks Soar Opposite Attacks: Strong Message to Terrorists?" Earlier, in a report about North Korea, they helpfully identified the nation discussed with the slogan "The North Korean Menace." As I exercised, I listened to the iPod.

I had recently deleted "Picture," a duet by Kid Rock and Sheryl Crow. In reading some shouty books by conservative authors, I had learned that Crow was quoted saying that war was never the answer. This has meant widespread ridicule of Crow and a place for her on the list of "loudmouth celebrities who should shut up about politics," a list that for conservatives never seems to include, well, all of the other people on my iPod. No one ever tells Lee Greenwood to shut up. I was sad to see Sheryl go. Her duet with Mr. Rock is a lovely song where she deals with a broken relationship by "heading to church" while Kid is "off to drink [her] away". Kid Rock refrains from his regular self-aggrandizing rants about "bitches" and his own testicles to mourn the loss of love that always seems to come from consistently committing acts of adultery while ingesting massive amounts of cocaine and whiskey. And again: he's the conservative, she's the liberal.

I had also deleted the Garth Brooks music I had loaded earlier. Something about it had bothered me for a while, but I couldn't put my finger on what. Then I remembered: Brooks's sister is a lesbian and in his song "We Shall Be Free," he puts forth the radical notion of being free to love whomever you choose. He didn't even say "marry," just said "love." Yes, that could possibly be construed as mere libertarianism. But no, it was not part of the conservative cosmology, and yes, the mere mention of tolerance, not even approval but mere tolerance, had infuriated conservatives. So yes, it had to go.

One song that had been in heavy rotation was Toby Keith and Willie Nelson's "Beer for My Horses." When I downloaded it, I figured it might be an amusing interplay about getting a horse drunk. I did not expect it to be a honky-tonk endorsement of preemptive lynching of anyone who could have potentially committed a crime or has the capacity to do so in the future.[2] When Willie and Toby's lynching-based crime prevention program is implemented, in the Brave New Toby-Willie World, the day's mission will be concluded with boozing. They intend to have whiskey for themselves and their men and serve beer to their horses. After "The Angry American (Courtesy of the Red, White, and Blue)," "Beer for my Horses" is one of Keith's most popular songs. If he and Willie ran for public office together and this song constituted the entirety of their campaign, I am certain they would get at least 35 percent of the vote in this country.

One of the Charlie Daniels songs I had on the iPod was his cover of the Lynyrd Skynyrd classic "Freebird." It's like Daniels took the original version, then took a fiddle, then shot you through the head with the fiddle, then lit the original version on fire. Then punched you in the kidneys.

My day of family life, neighborhood adventure, and exercise had not cured me of the creeping bastardness that seemed to be overtaking me like a cancer, but at least for a day it had not metastasized. I had been conservative that day, staying within my traditional world, tending to my family, and, even in the apolitical sense of "conservative," just not doing a whole lot. I had conserved. It had even been a relatively healthy day, but that's just because I had returned to what was already familiar to me in my house and hood, and therefore not out of my comfort zone and not really in keeping with The Experiment at all. Regardless of whether the day had aided the cause of conversion, my family liked me better and I had at least delayed any potential heart attack. My eye had barely twitched all day. As for comfort zones, the next day would find me solidly outside of mine.

---

1. Opposing fans call themselves Fans Against Gordon, an entirely intentional acronym. Gordon's preferred "rainbow" color schematic probably doesn't help matters.

2. I always figured Willie to be a fairly liberal guy, what with the ponytails and the pot smoking, but then again he did get into trouble for not paying taxes, which could have been purely motivated by his pressing endorsement of smaller government, so perhaps more libertarian.

# BRAZIL (1985)

**Summary:** In director Terry Gilliam's nightmare vision of the near future, the world is choking on its own bureaucracy, regulation, and mechanical apparatuses. The story centers on Sam Lowry, a low-level employee at the vast and sprawling Ministry of Information, who spends a lot of time daydreaming about breaking out of his dreary existence. When Sam spots Jill Layton, who looks exactly like the woman in his recurring fantasies, their lives become intertwined and Sam sets out to rectify the government's arrest of Jill's neighbor, apprehended because of a typographical error. Terrorism is a constant presence in this world (explosions rock a restaurant and customers go right on eating) and the prosecution of terrorism is the Ministry of Information's central mission. Eventually, as Sam slips in and out of daydreams centered on Jill and escapism, he is fingered as a terrorist himself and consumed by the state. Also, there are air ducts collapsing all over the place and Robert De Niro plays a renegade plumber/folk hero.

## CONSERVATIVE MESSAGES

- Heavy regulation, or perhaps socialism, means that you have to count on Central Services, specifically Bob Hoskins, to fix everything and they are as rude and unprofessional as the entire diverse plumbing and contracting industry is today. But in this dystopia, they have government contracts.
- When wasteful and lazy government employees at

the Ministry of Information, unmotivated by the competitive challenge of the private sector, are not being properly supervised, they watch old cowboy movies on television.

- The need to pay attention and be proactive about terrorism is demonstrated by the pervasiveness of terrorism throughout the film.
- Sam almost doesn't get to meet/help Jill because some dreary environmentalist keeps complaining about littering.
- De Niro's character operates outside of the system and is famous for his heroic brand of renegade capitalism. He is ultimately destroyed, literally, by paperwork. Which is exactly how trial lawyers like John Edwards would like it.

## ANTICONSERVATIVE MESSAGES

- We never know for sure if the terrorist acts are being committed by guerrilla groups or by the bloated government itself in order to better control the citizenry.
- The far-reaching and ineffective Ministry of Information, which shrinks the boundaries of personal liberty or often obliterates them all together, tends to make one think of the similarly broadly titled Department of Homeland Security and National Security Agency. Damn, Central Services reminds me of Halliburton.

**OVERALL PERSUASIVENESS SCORE: 81**

# 18

# The **Gay,**

## The Kangaroo, and the

# **Proselytizing** of

## Sammy **Sosa**

☞

**In which the author deviates briefly from the path
of strictly political conservatism to tread the path
of religious conservatism, in order to better
understand the threat of forces that the author
frankly never really saw as threats to begin with.**

In modern social conservatism, as I learned at the Family
Research Council a couple of weeks before, The Gay is flat-
out terrifying. A big, prancing, show-tune-belting menace.
Several dear friends of my wife's and mine were The Gay and
I've never been able to dislike them, much less hate them.
Sure, I felt a little left out when conversation turned to fash-
ion and Judy Garland, but that's not the same thing as hate. In
all my years of experience with The Gay, and I was a theater
major in college, The Gay had never once attempted to push
Gay on me. In recent years, as The Gay put forth the idea that
they would really like to be able to get legally married like The
Straight gets to do, I had no idea why that was being called a
threat to my marriage.

I didn't wish to hate or fear my friends, even for the purposes of The Experiment, but I wanted to try to understand the perceived threat because it seemed pivotal to today's conservatives. To try to grasp it, I chose the Antioch Bible Church, which had gained a lot of attention in recent months in the Seattle area for its vocal opposition to The Gay. Microsoft, a company based out here that sells software for computers, had dropped its support for a proposed state law that would protect The Gay from being discriminated against in employment, housing, and insurance, by adding "sexual orientation" to gender, marital status, mental or physical handicap, national origin, race, and religion in the law's wording. The pastor of the church, Ken Hutcherson, had bragged about being the one who convinced Microsoft to back off its support of the bill, a charge that the company denied, a denial that Hutcherson said was a lie.

Not only was the church very concerned about The Gay, it was a huge church, with one of the largest congregations in the area.

Of course, going there meant leaving the city of Seattle. Again. Test-drive an Escalade, go to the suburbs. Want to buy a Christian gift? Leave the city walls. So it went again as I set out to the suburb of Kirkland to visit the Antioch Baptist Church, which is a "church" in the sense of bringing people together to talk religion but not in the sense of being a building. Services are held in the rented gymnasium at Lake Washington High School. Between the two services held every Sunday, they draw about three thousand people per week.

The first thing you notice at the gym/church is how incredibly diverse the parishioners are. Plenty of white people, plenty of black people, lots of Asians, a few Latinos. It looked like an open casting call for a Benetton ad; the kind of diversity every liberal says they would love to achieve in their own community but can't quite pull off. The second thing you notice is how many kids there are. Everywhere. Apparently the Rexburg Idaho baby hose has been hooked up a few times here as well. Liberals take heed: conservatives are making a lot more of themselves than you guys are.

Before the service started, I figured I'd try to get some insight into what was so wrong with The Gay. I approached a young, stout, and bespectacled woman named Danielle who was sitting at a table soliciting donations to the youth program, told her it was my first time visiting the church and did she have time to answer a few questions about things? I said I had read in the newspaper about everything going on with Microsoft and the church. "How is it going with The Gay?" I asked.

"The what?" she asked.

"The Gay. You know with the church and The Gay." I half whispered, "Are you able to stop The Gay?"

"I don't think I understand."

"You know with The Gay and what The Gay wants to do. And the getting married to each other. How is it going trying to stop The Gay from doing that?"

"Well, there was the bill in the state legislature," she said, talking to me as a polite person would talk to an idiot. "It's a bill that's been going around for a while and it's antiharassment. Like sexual harassment and discrimination. And they wanted to put a clause in there about sexual orientation and we fought to stop that."

"And did it pass?" I leaned in a little closer and asked, "Did you stop The Gay?"

"I don't think the bill did pass this time, but it will be back next year. We're a Bible church. We believe in everything in the Bible and we believe that that's just wrong."

"The Gay, you mean?"

"And they want to force churches to hire people who don't believe in that church's doctrine and we believe that the Bible says it's wrong and we shouldn't have to hire people like that. It's like divorce. We don't hire people who are going through divorce because we don't believe in it."

"How about if you're injured? Like if you hurt your leg or your arm or something? Is that okay?"

"I'm not on the hiring committee."

I thanked Danielle for her time.

Entering the gymnasium where the service was to be held, I was dismayed to learn that Hutcherson was on vacation and would not be leading the proceedings. I was heartened to find that he was there anyway, hanging out and talking to people. If I was going to understand the threat posed by The Gay, I figured I might as well learn it from the man who either altered the policies of one of the most powerful corporations in the world or was so strong in his convictions that he deluded himself into thinking that he had.

Hutcherson is a former NFL linebacker and played on the very first Seattle Seahawks team in 1976. He showed up for drills the first day back then wearing a T-shirt saying HUTCH IS GOING TO SEATTLE TO DO GOD'S BATTLE because presumably other teams in the league were aligned with Satan that year.[1] I sidled up to him, looking for, uh, well, um . . . wisdom . . . ?

Suddenly he turned to me and grabbed me by the shoulder. His remark of "Hey, brother, how've you been?" seemed to indicate that he was not about to use his enormous size and strength advantage to kill me. He figured I was part of the congregation but that he couldn't remember my name.

"How's it going with The Gay? With stopping The Gay from marrying everybody?" I asked conspiratorially. Unlike Danielle in the lobby, he was not confused by the nomenclature.

"Pretty well, pretty well. I think we're winning. We don't go on our schedule; we go on God's schedule," he said, pointing up to where, I guess, God lived, "but I think it's moving in our direction."

That was all he said about the subject. I expected him to go off on a tirade, to viciously rip into The Gay and all the devious schemes The Gay was hatching. To explain to me clearly and succinctly what was so terrible about The Gay in the first place, but no, the church was on God's schedule, they were winning, and I was his brother.

The premise of Antioch Bible Church was a simple one and classically conservative: whatever is in the Bible is the law. While conservative Christians have long insisted that the Bible declares homosexuality to be wrong, liberals enjoy pointing out all the other things that it also says that don't really cleave to our contemporary society.

I told Hutcherson that I had these crazy liberal friends who say that the Bible tells us that we shouldn't eat shellfish and that we should sacrifice people. "What should I tell them when they start *talking* like that?"

"You just ask them *where* the Bible contradicts itself. Where does it contradict itself?" he asked me. I stammered because I didn't know (a pastor of a Bible-based church has a huge rhetorical advantage over me when it comes to Scripture). "It doesn't. And that stops them dead in their tracks." He said that yeah, there was stuff about sacrificing at the temple and not eating shellfish, but that was taken care of with God's death.

"*God's* death?" I asked. Wow, I thought, I'm pretty good about keeping up on the news, but I hadn't caught that one.

"When Jesus died," Hutcherson explained, "that was absolution for everything in the Old Testament. Then the New Testament came along and that was what we had to follow."

"So we're off the hook with all the sacrificing and we're free to eat all the shellfish we want?" Yes, he said, we were. And with that, he clubbed me hard on the shoulder and left my company. A small and simple bruise soon formed.[2]

Because Antioch Bible Church conducts its services in the gymnasium of Lake Washington High School, they don't have to worry about launching a capital campaign for a building, so they're not dependent on fluctuating tithing as the economy goes up and down. This gives the church some economic flexibility but also presents some challenges in terms of art direction because the mascot of Lake Washington High School is a kangaroo. So as the pastor and other guests speak, they are surrounded by big purple banners proclaiming love and allegiance not to Jesus but to a kangaroo. It's like some sort of neopagan Australian idolatry.

I was listening to the service, trying to catch some mention of The Gay, but there wasn't any. Some people who were missionaries in the untamed heathen jungle of Belgium were sharing their story and some hymns were being offered. The lyrics to the hymns were projected on a big screen, but my view was blocked by a not-fully-retracted basketball hoop, and being a godless Seattleite, I didn't know the words by heart. So I started to drift into this weird reverie, musing on the notion of Jesus being a kangaroo who played basketball and we were all here to watch His big game against the Devil and a team of The Gay. Our Lord and Savior Marsupial point guard! Do the "hide the ball in your pouch" trick! I decided it might be a good time to go get a little bit of air. Since I was seated on high school gym rafters instead of a pew, it was no problem getting up and out of there.

I was going to walk around to the other side where I could assess things a little easier and where there was more foot traffic. As I came around the building I was apprehended, spiritually speaking, by Sammy Sosa. Not literally the bulging major league slugger but a five-foot seven-inch slimmed-down version[3] who had overheard me in the lobby earlier talking to Danielle. I think he was suspicious of my use of "The Gay" in that conversation and wondered if I was up to something.

Sosa asked in a friendly way why I was there and I told him I just wanted to learn.[4] A wide-ranging discussion of spiritual matters followed as he explained to me his beliefs and the simplicity of it all being in the Bible. We talked about how he knew the word of God was in the Bible and not some other book like *The Iliad* or *The Odyssey*. It's God, he said, because it was written by sixty-six people over thousands of years and there are no contradictions in it. Part of me knew this could not possibly be true,[5] but he had done his homework on this and I was rhetorically cornered. I was unable to cite Scripture and verse to form any counterargument.

So I went secular and asked him about the science of the Bible. Scientists tell us that the earth is billions of years old and life is millions of years old, but the Bible says that it was all created in six days. I wasn't trying to debate the existence of God, or the value of the Bible, I was only hoping that by tracking his thought process I could eventually find the path that led to the part about being mad at The Gay. Sosa told me that he has struggled with the time thing himself, since he comes from an engineering background. Still, he said that since time was a concept created by man, everything was relative. Since God was all-powerful, He could have created an old earth. "Really?" I asked him. "An old earth? Like He made all the rock formations and planted the fossils in there and everything?"

"Well, that's one theory. There are lots of different theories. I'm still working on that one myself," he admitted.

Ten minutes later, still nothing about The Gay. I moved away from science and to more contemporary issues. Danielle in the lobby had talked about the church's opposition to divorce and I asked Sosa for clarification on that. People shouldn't get divorced, he says, because marriage is a covenant you make with God. That's the covenant that the Family Research Council woman had talked about, the one the state should make it harder to rend asunder.

"What about if the husband is beating up the wife and taking her money and being a danger to her safety?" I asked. Sosa said that safety is a consideration and that they would try to get both husband and wife to a safe location and try to minister to each of them from there in order to bring them back together, but the vows of marriage, he said, don't say "till death do us part UNLESS someone beats someone up or takes their money." Some people end up getting divorced anyway and at that point the church considers them adulterous and they're kicked out. They believe you're supposed to stay married to the first person you marry. Forever. Period.

Religious conservatives align neatly with political conservatives because they believe in instruction manuals. Got a court case? Read the Constitution. Spiritual problem? Look it up in the noncontradictory Bible, especially the shellfish-sanctifying New Testament. The instructions, like well-crafted technical writing, are apparently clearly spelled out.

It was all very interesting and possibly illuminating at some later point, but my feelings on the threat that same-sex marriage posed to my own marriage had not changed. I forced the issue. "That's all fine. But let's talk about The Gay. What's so bad about The Gay?"

Sosa paused and took a breath. Antioch Bible Church had received a lot of attention on the whole The Gay thing and carefully considered what to say about the issue. Sosa said the Microsoft action got a lot of publicity and a few protesters showed up at the church but the church doesn't really talk about homosexuality all that much. Yeah, I *know*, I said, demonstrably annoyed by this point. The church got involved with Microsoft, their neighbor in Redmond, because they felt like Microsoft, by supporting the bill, was trying to push its idea of homosexuality on the rest of society. "God doesn't differentiate between sins. It doesn't matter if its murder or adultery or homosexuality or anything else. But if you are saved and you confess those sins and repent, you still go to heaven."

"So you can run around and gay all over the place," I said,[6] "but if you're born again before you die then its okay?"

"Well, it's not okay, but you still get to go to heaven."

"Love the sinner, hate the sin?" I asked.

"See, people use that phrase a lot and I think it oversimplifies things," he said.

"So 'get really frustrated and annoyed by the sinner and also hate the sin'?" I offered.

"Yeah, that's right."

Our conversation tapered off after Sammy asked if I worried about what would happen if I died tomorrow and I said, nah, the life insurance is paid up, so the wife and kids will be fine. Well, he asked, had I been saved?

Hmm . . . I thought for a minute. "Nope. Not really. No, can't say that I have."

We were about to move into a new phase of the dialogue when I cheerfully told him that I had to go. I had taken what I needed from the church and was still working on the idea of murder, adultery, and gay being equal offenses. I guess it was a little much to expect to be converted to that way of thinking over the course of one Sunday morning at a kangaroo shrine. I never did learn how The Gay had endangered my marriage. So either they weren't actually endangering it at all or—OR!—they had hatched a stealthy plot, the full extent of which I was too naive to grasp. Of course, one could credibly argue that perhaps the greater threat was created by my going over to a gymnasium church service of the type my loving spouse and liberal children would want no part of. So I drove home. I was flying to California the next day.

---

1. I believe Medved played for one of those NFL devil teams. A punter, I think. Can't confirm that.
2. Legal charges are pending. (No they're not. But ow.)
3. Like how he looked before all the steroids. Alleged steroids, I mean. Because maybe the real Sosa just quickly became naturally, organically really really huge for no apparent reason.
4. True, I might add.
5. I checked later. There are skazillions of people pointing out contradictions, and skazillions of other people saying there aren't any contradictions.
6. Marking my first instance of *gay* as a verb.

# 19

# Twinkle, or Man,
## That Ronald Reagan
## Sure Loved
## Popcorn

☞

**In which the author journeys to Simi Valley,
which is that place where the cops who beat
the crap out of Rodney King were acquitted, which
was a verdict that caused Los Angeles to, you know,
more or less explode. That's where the author
goes to learn about "The Gipper."**

You can learn a lot about your kid by letting him choose what book he wants for a bedtime story. You can learn a lot more about him by noticing which book he wants to read with mind-numbing persistence. As we lay down that night, Charlie chose his longtime favorite: the Dr. Seuss classic *The Lorax*. We'd probably read the book over three hundred times in the past two years. Some nights twice. I know that it takes me precisely twelve minutes and twenty seconds to read aloud. For a good six months when he was two, he insisted on it every night. I could read it with all the voices and dramatic inflection but still be able to plan my schedule for the rest of the week, think about how John Edwards fills all the time he now has on his hands, whatever. The premise of the book is this: the

Onceler is an industrialist who moves to a forest full of truffula trees. He realizes he can chop them down and turn them into "thneeds," which are big floppy all-purpose garments that double as bicycle seats. A small "kind of a man," the Lorax, shows up and serves as the de facto Ralph Nader of the story, warning about the dire consequences of the Onceler's plan. Nonetheless, everything works fine in the unregulated free-trade economy of Seuss and the Onceler is making a fortune selling the things, that is until all the animals nearby get sick or leave as the forest is denuded. Finally, there are no more trees and the forest is a barren landscape and the Onceler is out of business and despondently remorseful over what he's done. Unlike Ralph Nader ever has or ever will, the Lorax leaves without saying a word.

It wasn't until the end of the book, when the forest of truffula trees has been devastated by the Onceler's capitalist ambition, that I realized I was reading something that would clearly be off the list of acceptable materials to expose myself to. Fine for Sierra Club son, certainly not fine for Dad.

"Charlie, have you ever thought that maybe the Onceler had good reasons for chopping down the forest?" I asked him.

"No!" my son said firmly, then paused to stare at me for a second before reiterating, "Dad, no!"

"But he wanted to make money. Sure, things got away from him a little bit, but remember when he hired all his uncles and aunts to knit the thneeds? He was creating jobs and making money and providing thneeds, which, as it says in the book, everyone needs."

"Dad! He cut down the whole forest! That was not nice!"

When my son and Dr. Seuss team up together, I'm never going to win the argument.

I didn't see Charlie the next morning because I was on an early flight to Los Angeles. After a couple of weeks immersed in the cultural side of conservatism, I was ready to wade back into politics. In so doing, I would visit a monument to a man who served as a totem for both: Ronald Reagan, the patriarch of modern conservatism. He swayed Rich Lowry, and if you are to believe the people I had been reading and listening to for the past three weeks, he had defeated the Soviet Union pretty much by himself. I was heading for Simi Valley, home of the Reagan museum as well as the jury in the Rodney King case, as well as some excellent wine. As long as I was in the Los Angeles area, I planned to take in the Nixon museum as well.

On the plane, I pulled up some Kid Rock music and pulled out Sean Hannity's book *Deliver Us from Evil*. In Bill Kristol, Rich Lowry, and Jonah Goldberg, I had met smart conservatives. In Mayor Shawn Larsen of Rexburg, I had met a down-to-earth practical conservative. In Michael Medved, I had even met conservative talk-show host with whom I could have a civil and enlightening conversation. While I didn't agree with everything they had said, I found something resonant in each of them. Hannity? Not so much.

Hannity's vernacular is as objective as his network's. It's never "the *Washington Post*," it's always "the liberal *Washington Post*." And Lou Cannon didn't "say" something when he criticized Reagan, he "sneered" it. The *New York Times* is deemed "incorrigible" for criticizing Reagan. I always thought that you wanted a newspaper to be incorrigible. When I think of corrigible newspapers, I think of *Pravda*.

Likewise, according to Hannity, the shah of Iran had "human rights problems" and "wasn't a perfect ruler." Well no, running a murderous secret police program is rarely considered the mark of perfection. Frustrated, I put Hannity away and spent the rest of the flight with a golf magazine and a copy of *BusinessWeek* I was fortunate enough to find on board.

There was a lot to see at the Reagan museum, I figured, and

I planned to dedicate an entire day to absorbing it. I was going to need some stamina if I was going to pull that off. Other conservatives had been swayed over the course of a full Reagan campaign or a full term of his presidency. I had to get it all done in a day. Complicating matters, I was in a pinch with the food I had consumed. There had been a small scone on the airplane and I had worked through parts of three cups of coffee (7-Eleven coffee at 4 A.M. on the way to the airport, a twelve-ounce Starbucks at the airport at five-thirty, part of a watery cup of ick on the plane at eight). Regardless I was starting to get awfully sleepy. Like, not able to concentrate at all, head full of fuzz, in really bad shape, veering off the freeway during L.A. morning rush hour. Fortunately, when I stopped at Office Depot for some tape-recorder batteries, I noticed they had beef jerky. It was like Popeye's spinach. Got it down my throat and my mental muscle popped out. It was all "daa-da-da-da-da-da-daaa!"[1] I was ready to kick some ass.

To get to the museum, you have to drive out of Los Angeles, into Simi Valley, and along a long winding road that takes you up the side of a big hill until you finally arrive at the massive Reagan compound. It has a sprawling view of huge land tracts where new homes will be built over what may have once been vineyards, and it exists as an island unto itself, pleasant and detached from any other businesses, homes, or conventional reality known to other people. Again, this is the *Reagan* museum we're talking about. If the numerous highway signs that clearly spelled it out didn't tip you off that you were in Reagan country, the bronze Reagan statue out front solidifies it. It's enormous, standing eight feet tall, looking like a cross between Reagan and NBA legend Gheorghe Muresan.

If I was going to be staying until the 5 P.M. closing time, I would be underfed and not sufficiently caffeinated since they did not appear to have a café or even vending machines.

I didn't have time to go all the way back into Simi Valley for sustenance, so I went to the gift shop and bought a large bag of Jelly Bellies. The goo would have to serve as food while the sugar would be sitting in for caffeine. Besides, what better way to get to know Ronald Reagan than being somewhat sleepy, confused, and high on jelly beans?

I was assigned Gail, a small kindly elderly volunteer guide with curly hair and glasses, and after I paid full-price admission (there are no handouts in Reagan country, at least not for people in my income bracket), we went on a tour.

One of the first rooms you come to in the exhibit is a hall of portraits featuring all the presidents, most painted by artist Larry Williams, who has made each chief executive look like a jowly and incandescent cartoon character. Williams died somewhere along the way, so he never got past Reagan himself, leaving Bush the elder, Clinton, and President George W. Bush to be represented by spooky photographs, chosen by the presidents themselves. The first Bush chose a sort of photo illustration and Clinton opted for a really fuzzy out-of-focus thing where he looks like he's falling over. As for the present Bush, he chose a picture that makes his head look big. Like enormous. Hydrocephalic big. While every other president is pictured from about the belly button up, President George W. Bush is seen from the bottom of his necktie knot to the top of his head. He looks like he must be nine feet tall and weigh five hundred pounds. What's the deal with enormous Republican presidents in this place? I wondered.

For all of the cowboy imagery and the horses and the California governorship, Reagan was from the Midwest. He was born in Tampico, Illinois, in 1911 and moved to Dixon, Illinois, when he was nine years old. During his tenure as a lifeguard there, he is credited with saving seventy-seven lives. This fact amazed me. I realize people's legendary status gets puffed up if there's a museum dedicated to them, but come on. I've swum my whole life and have seen a life-guard save someone maybe once. Maybe. "Were the people of Dixon exceptionally lousy swimmers?" I asked Gail. She said she thinks many of those Reagan was credited with rescuing were young women who didn't really need saving but wanted Reagan to pull them from the water anyway and possibly give them mouth-to-mouth. Though shriveled and spooky-looking later in life, he was pretty handsome as a youngster. My mother, who was a girl during Reagan's film-making heyday, used to have a crush on Ronald Reagan, a revelation that would later disturb her and me.[2]

During and immediately after college, Reagan worked in radio, broadcasting the Cubs and sometimes the White Sox games from Iowa, relying on wire reports and then describing them extemporaneously, pretending he was watching the game while imagining it in his head. Part of me really liked that. As a radio guy myself, I sometimes have to improvise, especially during pledge drives. I made an effort to be charmed by Reagan's improv skill and not be troubled by this early instance of mutable truth in his career.

The story goes that while he was covering Cubs spring training in California, one of the games was rained out. So Reagan went to visit a friend in Hollywood, got a screen test, went back to Iowa, and two weeks later got a telegram with an offer of a $200-a-week, five-year deal to be a "contract player" with Warner Brothers. The "studio system" was still in effect, meaning actors essentially belonged to a certain studio, performing whatever roles the executives told them to play and saying whatever was in the script handed to them by a consortium of influential figures operating behind the scenes. Again, this is *Ronald Reagan* we're talking about.

Puppet imagery aside, I was starting to like Reagan because

I was identifying with him. In my own performing career, I had gone in the opposite direction of Reagan, moving from acting into radio. Maybe I'll become a lifeguard next. And then a baby in Tampico, Illinois.

A rising star, Reagan enlisted in the army and was immediately assigned to make movies for them. "Propaganda films!" Gail said cheerfully. After the war he went right back to making movies, clips of which were shown at the museum in a special tiny screening tent sponsored by Warner Brothers. Maybe it's the clips that they selected, or the couple of hours I had already spent in a cult of personality shrine to the man, but you know what? He wasn't a bad actor. He was kind of like *Good Will Hunting*–era Ben Affleck crossed with, like, Noah Wyle. Lefties always call him a "second-rate actor" or a "B-movie actor," but he's really pretty funny and sharp delivering the rapid-fire exposition found so often in Warner Brothers movies of the day. Even in *King's Row,* a dramatic lead where he's in the highly strange circumstance of having his legs amputated because an evil surgeon hates him,[3] he plays it with, you guessed it, simplicity. He leaves himself alone.

This is what I realized about Reagan watching him in that performance, and then thinking about his presidency. He left himself alone. "Leaving yourself alone" is an acting term where you exist solely in that moment on the stage, listening and responding to whatever is happening right there. The opposite of this is "being in your head," where you're worrying about your performance, comparing it to previous performances, worrying about upcoming moments, and doing just about anything rather than simply being in the moment. My acting career was doomed because I was constantly in my head. Reagan left himself alone and it served him well.

"And this is when he was testifying at the House Committee for Un-American Activities. And over here are two ships he put together when he was waiting between takes in one of his movies." Whoa, whoa. Back up.

"He testified in the McCarthy hearings?" I asked.

Gail pointed to a Lucite box with some transcripts of Reagan's testimony and a picture of a bespectacled Reagan, then president of the Screen Actors Guild, in a rare nonsmiling shot. According to the display, Reagan was no great fan of communism but had it under control within the union and believed that through our regular democratic system, the threat would pass. Even with Reagan blinders on, I knew that McCarthyism, communism, and the 1950s were a hell of a lot more complex than that. Reports indicate that Reagan was giving the names of suspected communists to the FBI during his tenure at SAG. The Cold War was looming, no one knew what to make of the world being formed, and the threat of communism freaked folks out. But the Reagan way of looking at it, or at least the way Reagan employees had of presenting the way Reagan looked at it, was just to keep being chipper and wait for goodness to sweep over the earth once more.

Leaving the union and the screen, Reagan moved on to elected office. If only politics itself were as easy as the path to greatness that was illuminated in the displays of Reagan's ascendancy to first the California governor's mansion. "This is the size of the desk he used when he was governor of California," said Gail. "He was a very popular governor. He got a lot of people off welfare who shouldn't be on welfare. They were double-dipping!" Then it was on to the White House. Brief mention is made of the failed 1976 run[4] but it's pretty much a clear road to the White House in '80 and '84 according to the small panel of buttons used to signify campaigning. No information given on why he ran, what positions he ran on, or really what he stood for. Just a few perky buttons and yard signs, and *boom*. He's there. President. Meanwhile, the internal workings of the Reagan presidency, the eight-year period that all these conservatives seemed to be pointing at as the definition of When Things Were Right with America occupies shockingly little space at the museum.

In one enclave, I watched a video montage of the missile programs that were important in Reagan's day. There was a small panel installation of the SDI or so-called Star Wars program. "This was to protect our country," said Gail hopefully. "The missiles would be shooted [sic] down before they . . . they . . . they . . . got here." What I knew of the program was that it was insanely expensive and if it had ever been built it would probably not have worked. I didn't say that out loud.

Another video montage told the story of Reagan firing the air traffic controllers after they went on strike. Gail told me that Gorbachev saw how those events unfolded and that filled him with terror about dealing with his tough American counterpart. Gail's visualization, dramatic though it was, seemed a bit wobbly given that Gorbachev would not be leading the Soviet Union for another four years.

In one corner, there was a small looping video presentation about Reagan getting shot. Being shot does not fall under the rubric of "things accomplished in office," but I admit it takes a measure of moxie to have a museum dedicated to one person and then a video of that person getting shot and looking perplexed over and over. A father and his seven-year-old son had come to the museum and the dad called the son over to see the video. "Look, this is from when someone shot him," he said.

"What movie was that in?" the son asked.

"See, he comes out of the hotel and they shot him."

"Which one was this from?" asked the confused son. He never got his answer. The red sweater that Reagan wore home from the hospital is encased in glass.

There had been displays of Reagan as a lifeguard, pictures

of Reagan's boyhood home, letters between Reagan and second wife Nancy, even a container of popcorn to denote the favorite snack of his youth. Maybe the museum was simply choosing to observe his entire life and not just his years in office, but, you know, lots of people eat popcorn. Few become president and fewer still have the apparent impact of Reagan. I was frustrated because I felt like I needed an impressive, masterfully arranged display, possibly with lasers and a fog machine, of those eight years in order to get converted. I hoped I would find more about Reagan's policies when Gail took me through a door and down a ramp toward more displays. Soon we were in a re-creation of the Oval Office. The floor used to be level with the previous rooms and the ones still to come, but when Reagan was originally shown the Oval Office room, he complained that the ceilings were not as high as the real thing. So they dug a pit, dropped the whole Oval Office down, and put in higher ceilings. I guess that's something to admire about Reagan, I thought. I can never even get a contractor to put in a working light switch.

The Oval Office was arranged carefully and featured jars of Jelly Bellies and pictures of FDR and Eisenhower, both of whom Reagan admired.[5] After looking at the office for only a couple of minutes, Gail moved us along to a large room dedicated to Reagan's vacation homes and other leftover pleasantness from his presidency and later life. The nuts-and-bolts parts of the presidency were over. No explanation of tax cuts and trickle-down economics. Nothing on social policy. Not even anything about the Libya bombings, which I remember during high school as an unnerving "Am I going to have to go fight this war?" moment in my life.

Popcorn wasn't enough. I wanted to be inspired by the Reagan revolution like Rich Lowry had been, but if it had not happened in the previous rooms, it had to happen in this last one. I furrowed my brow as I tried to really suck in the majesty of Nancy's inaugural gowns. I looked really closely at the autographed ball signed by Adrian Dantley and Vinnie Johnson of the Detroit Pistons. Nothing happened. There was a framed cover of *New York* magazine that showed Nancy dancing with a smiling Frank Sinatra while Ronald Reagan tried to break them apart. Wonder how that got in to the museum. None of it made me a conservative.

In fact, there were sections that were counterproductive to the cause. Gail pointed me to a display of an appointment calendar. "This is his actual calendar for June 1988, right to the minute," she said. There were a fair number of items on his agenda but most were in the late morning to midafternoon, generally nothing in the evenings, and practically nothing on the weekends. For the president. Of the United States. Of America. It appeared he was working for us five days a week. Maybe six hours a day.

I flashed for a moment back to Richard Clarke's book about terrorism, *Against All Enemies.* Clarke, a former State Department official in several administrations before being squeezed out by President George W. Bush, recalled Bill Clinton stopping him in the hallway and complimenting him on a speech Clarke had given in Philadelphia. Clarke wondered how Clinton could have seen that speech and the president said he had watched it on C-SPAN. When Clarke checked the listings, he saw that C-SPAN had run the speech at two in the morning. I tried to stop thinking about that. Reagan knocking off well before dinnertime, Clinton enraptured by policy at two in the morning, President George W. Bush is famously in bed by ten every night.

I was well into the afternoon at this point and all I had had to eat was that sack of Jelly Bellies. As visions of appointment calendars, Reagan chopping wood at the ranch, and a "Reagan

1" jersey from the Los Angeles Lakers swirled around, I realized I was high. After several hours of Reagan reverence and only sugar and residual beef-jerky fragments in my system, I was looking at the eight-foot Reagan and imagining it coming to life and smashing the Berlin Wall to bits with its fists. "Reagan tear down wall!" it would shout. "Freedom gooooood!!"

My tour with Gail had wrapped up and I thanked her for her time. Before we parted, I asked her how Reagan would do as president today. "I think he would do well. I think he would bring people together." She also believed that the war Reagan led us through, some say resolved, some say won, was better managed than the current one being waged by the current president. "In the Cold War, it was expensive and no lives were lost. And now we have a war that's expensive and lives are being lost."

"Oh," I said, "wow. Okay."

Gail encouraged me to check out the rest of the grounds, including the scale-model replica of the White House south lawn, donated by Merv Griffin. It was indeed impressive, but more exciting was the fact that it turned out there really was somewhere to eat: the Reagan Country Café. I gasped with joy and ran inside. I considered getting the Air Force One Burger but was instead drawn to the F-14 Dog, a wiener as massive as the enormous deficits run up during the Reagan administration. I was pretty hungry, so I also got an order of Stealth Fries, which despite their name I could see plainly on the plate. With a couple more hours to go until the museum closed, I plotted how to maximize my time and give myself the best shot at Reagan conversion. Since Reagan loyalty was the bedrock of modern conservatism, I might finally be won over if I could get that loyalty to take place within me. Still, I wasn't feeling what all those converted "Reagan Democrats" must have experienced. I was staring at the same thing everyone else was seeing, but the picture wasn't coming in. It was ideological Magic Eye.

I know a lot of people put a lot of time and money into making the museum a fitting tribute to this icon of conservatism, but this one had failed to explicate what made him so great while in office. Seemed like a nice guy, don't get me wrong, with the lifeguard work and the movies and the popcorn and playing for the Lakers and everything, but where was the section that showed me why the tax cuts were such a great idea? And where was the depiction of the parts that didn't go so well? Granted, there was the morbid Hinckley-assassination-attempt tape loop, but how hard would it have been to install an Iran-contra display with animatronic Oliver North and Admiral John Poindexter? Give them big muscles and Power Ranger costumes if you want, I don't care, just show me something.

Looking for inspiration, I returned to the gift shop. There was plenty to choose from along with loads of elderly tourists as well as families towing kids with that "this is the stupidest place you've ever brought us" look in their eyes. Reagan dolls, Reagan puzzles, Reagan-embossed golf clubs and cowboy hats, and clothing with a logo of an American flag but with the initials *RR* where the stars would ordinarily be (so it looked like the country was composed not of the fifty states but of Reagan himself). My favorite item was the PEACE THROUGH STRENGTH T-shirts housed in a cardboard retail display case with the name of the shirts' manufacturer in big letters on the side. The manufacturer or distributor's name was "Yikes!" So you'd see the PEACE THROUGH STRENGTH T-shirt and then go around to the side to see "Yikes!"

The book section had plenty of Reagan bio/hagiography along with books about Eisenhower, the founding fathers, plus Donald Rumsfeld and conservative MSNBC talk show host Joe

Scarborough. Nothing from any modern Democrats but also no books about Gerald Ford. Granted, it would be hard to write a book about Ford, but there are some, and given his role in Reagan's political career, you would think there would be some representation. Also missing from these shelves: Richard Nixon, who was a Republican president, and President George W. Bush, who was mired in pretty low poll numbers at the time but was still, you know, the, uh, you know, president.

The bookshelf was a microcosm of the museum and the Reagan presidency. Everything unpleasant, be it Gerald Ford, Iran-contra, or, evidently, the current president who had opposed Nancy Reagan on stem cell research, did not fit the narrative and so was simply not part of the cosmology. It was a world as seen on horseback. In pictures, Reagan looked a little confused at birth, at the McCarthy hearings, and upon being shot, but other than that it was all smiles and twinkles. This was all being shown at a museum located far above Simi Valley, away from the complex urban universe of Los Angeles. The real world might as well be Jane Wyman for how divorced the Reagan empire was from it.

I did one last pass through the displays but tried to switch up my expectations and perspectives a bit. I slackened my jaw, widened my eyes, and attempted to convince myself, if only a little, that I already believed in everything Reagan stood for. Maybe by faking a belief, just a smidge, it could be like having ideological kindling that once sparked could create a roaring fire of conservatism. But the harder I tried, the less it worked. By trying not to think about all the mental patients released from in-patient facilities because of Reagan's domestic program cuts and all the people who died in wars in Central America, I only thought about them more while rereading Reagan's letters to Nancy.

Then again, maybe all the things I was struggling with were the very things Lowry and others loved about Reagan. The short hours, the way the presidency, which was a function of government, was performed as almost a hobby. By thinking that a president's job ought to be, you know, time-consuming, maybe I was worrying too much. I tried harder to adopt the blissful stance of someone who thought that the world really was this simple.

My giddy expression finally caught the attention of one of the elderly volunteer docents, who asked me what my favorite part of the museum had been. Hard question. I settled on the tiny McCarthy hearings display. Not because I liked it, just because it had the most drama to it, narrowly edging the Hinckley tape loop. I asked her if she knew any more about Reagan's involvement at the hearings.

"When Congress had those hearings, they called him to testify because he was president of the Screen Actors Guild. What he said was that there were some communists trying to infiltrate, but it wasn't as widespread as they indicated."

I was pleased that McCarthy, no matter how much Ann Coulter still stood by him, had been knocked down here. "Yeah, people got a little . . ."

"A little crazy," she said. "But I didn't realize until I started working here that if they infiltrated, they could start making movies, and you know you'd have . . ."

"Communists all over the place?" I offered.

"And a lot of them are creeping up again," she whispered, leaning into me.

"Really?"

"These rabble-rousers are being basically that," she said, leaning still closer, in case they were listening. "These protesters."

"The war protesters are communists?" I asked.

"Well, some people think that. But you don't know what their official position is."

"They might be sneaky?" I said, as sort of a halfway question.

"That's probably why he worked so hard to defeat them."

"The rabble-rousers?"

"Right."

My eye twitched. The nervous tic was back.

Back in the car I was scanning for Rush when I came across some news. Karl Rove, the president's political adviser, Mr. Therapy and Understanding, had been implicated in the Valerie Plame scandal. Indications were that he may have leaked, or approved a leak of, Plame's identity as a CIA operative to columnist Bob Novak and others as retribution against Plame's husband, former ambassador Joseph Wilson. Wilson had asserted in an op-ed in the *New York Times* that there was no credible evidence that Saddam Hussein had attempted to purchase yellow-cake uranium from Niger, which ran counter to the administration's case for going to war. I figured since the story had broken over the course of the day, and it involved the possible criminal prosecution of one of the most powerful people in the world, it might be something people talk about a bit.

But nothing. The host, someone local whom I never was able to identify between the blaring commercials that interrupted him every few minutes, offered plenty of information about how Islam is bent on world domination and how England brought the attack on itself by having loose and lax immigration policies, but nothing on the president's top political adviser being tied to a potentially criminal act.

Back at the hotel, I tried to unwind with a little TV. I have only basic cable at my house and can't remember the last time

I had any kind of movie channel. So staying in a hotel is extra exciting for me because I get to watch good movies without commercials. Maybe one of them newfangled TV shows like *Curb Your Sopranos* or *Sex Six Feet Under the City* would be on. Nope. They were showing *Fahrenheit 9/11*. Banned from The Experiment. C-SPAN was showing Bush speaking at Quantico, Virginia, so I watched that. Lots of platitudes, lots of smiling, friendly, I couldn't remember a word of it five minutes after it was over. Like Reagan but without the "inspiring a devoted following" part.

Maybe the Nixon Museum would work out better, I thought, drifting off to sleep.

---

1. Have you ever noticed Michael Moore's passing resemblance to Bluto?
2. And my father.
3. ?!
4. My love for Gerald Ford, dormant since the third grade when my dad supported him and I was blindly loyal to my dad, started to make a bit of a comeback here since the guy was given so little regard.
5. Remember FDR was a Democrat yet ass-kicky enough to be beloved by neoconservatives.

# 20

# Dicky

## Really Is

# Tricky

☞

**In which the author, who was quite young at the time of Watergate and never really understood it all that much when Nixon was resigning and waving from the helicopter and all that, tries to learn the truth about the most caricatured president in history.**

I woke up in my motel room in Orange County, a reliably Republican region in an overwhelmingly Democratic state. I downed a few glasses of tap water and, for the hell of it, stopped off for some orange juice, what with the name of the county and all. Then I set out for the Richard Nixon Library & Birthplace. I didn't shave before I left so I could try to rock that whole 1960 debate look. Didn't own a drab gray suit, but you do what you can.

The Reagan site had been pretty easy to locate. You head up north on 405 and look for the huge sign that reads RONALD REAGAN HIGHWAY. Then look for additional signs, located miles before they were really necessary, telling you which exit to take. Next, let several more easily noticed indicators guide you through the well-funded Simi Valley neighborhoods to the mountaintop palace known as Reagan Country or Reagan Planet or something like that.

Nixon? Not so simple. I followed Imperial Highway into
Yorba Linda, figuring there would be some signage indicating
the historical location and tribute to the town's favorite son.
But no signs. Anywhere. I know most modern conservatives
like to think that their movement began the day Reagan was
inaugurated, but guys, come on, Nixon was elected president
twice and had served two terms as Eisenhower's VP before that.
Would it kill you to put up a little "Turn Here" indicator?

The thing is, they ought to be proud of the Richard Nixon
Library and Birthplace, at least architecturally. It looks great.
A sprawling complex of white buildings paying tribute to the
life of the man, and, nestled in the middle of it all, the actual
house Nixon was born in. The Reagan facility had been jam-
packed on Monday, but there was hardly anyone at the Nixon
on this Tuesday morning. Just a handful of senior citizens, a
vanload of recalcitrant field-trip teens, and me. When I had
inquired a few weeks before about arranging a private tour
of the museum, I received a phone call saying absolutely, they
would be delighted to give me a VIP docent-led guided tour,
much like the one I had received at the Reagan museum, but
when I actually showed up, there was confusion at the front
desk and a lot of incredulous animosity coming my way from
Beverly the admissions/gift-shop cashier on duty.

Beverly, pinched, elderly, and resembling the meanest school
lunch lady you ever had to deal with, sputtered and ranted
about the impossibility of my presumptuous plan and called
the administrative offices to, I believe, have me slain. Soon,
two employees came down to move me away from Hurricane
Beverly and explain that they couldn't give a private tour that
day but would gladly give me free admission and point me
to the various volunteer docents. "Thank you," I said, eyeing
Beverly icily. So much obstruction and cover-up! Did I mention that this was the *Nixon* museum?

As there was at the Reagan museum, a film was played at the Nixon facility that gave an overview of the president. Unlike the Reagan, this one was long, in-depth, and didn't gloss over scandal, conflict, and difficulty. Watergate, turmoil over the war in Vietnam, and other challenging and unpleasant parts of the Nixon years were discussed, albeit in much more positive tones than Nixon historically receives. The film also had a poignant quality, talking about how Nixon was feeling better than ever and was making a big comeback. The film was produced before the 1994 death of Richard Nixon but remained, hauntingly, in constant rotation eleven years later.

As if to let you know that they realize the guy's actually dead, the exhibits open with a look at Nixon's funeral coverage. Then we jump straight to Nixon being home from the war and running for Congress. No boyhood, no lifeguard work, just presto, thirty-three years old. He wins. Within a few feet of carpeting, he's already running for Senate and defeating Helen Gahagan Douglas, a congresswoman and daughter of actor Melvyn Douglas, and, according to Nixon's campaign at the time, a damn commie. "Now she was a *liberal,*" said the seventy-year-old gentleman docent to the group of teens he was saddled with, "and do you know what that means? It means you liked everything the communists stood for and were probably a communist too." The surly teens paid no attention.

It's easy to kind of chuckle at the idea of communism today. With the collapse of the Soviet Union and the changes in Eastern Europe over the last twenty years, communism is now mostly found in freshman college courses, the idealism of college freshmen taking those courses, and obscure candidates, located at the back of the voter's guide, who don't even have Web sites. Communism is history. But in the forties and fifties, when a scrappy yet jowly Dick Nixon was coming up through the ranks, communism was current events. An investigation of

Alger Hiss conducted by brash young lawmaker Dick Nixon led to the veteran diplomat and accused communist Hiss being convicted of perjury and sentenced to prison. The notoriety of that case greased the way for Nixon to become vice president under Eisenhower. Back in the days before Clinton-Gore and Bush-Cheney, it was entirely unnecessary for presidential running mates to be pals, have family connections, or even know each other at all. With that freedom plus a contrast in age, geographical region, experience, and hair, along with his new-found reputation as a red-busting badass, Nixon got the nod. They won and Nixon held the job for eight years.

I remembered an old anecdote where someone asked Eisenhower what Nixon's greatest accomplishment was as vice president and he said, "If you give me a week, I might think of one." Everyone chalks that up to Nixon being without accomplishment (and Ike being kind of a dick to Dick). I think it might have been because Eisenhower never saw Nixon. Judging by the photos on display, it looks like all Nixon did was travel. Fifty-six countries in eight years. Not much indication of him actually accomplishing anything there outside of having what looks like a pretty great time, but he went there and that counts for plenty. What also counts are the times he was not having fun at all, such as on trips to Peru, where he kicked an aggressive heckler in the shins, and in Moscow, where he angrily squared off against Khrushchev at a display of futuristic home appliances in what was called "the kitchen debate." Can you imagine any modern vice president doing stuff like that? Mondale? Gore? Cheney might want to kick someone in the shins, but what are the odds that his body could tolerate the exertion without it leading to a stroke?

Americans loved Nixon and things were going great.

Then they stopped going great.

A display at the museum maps out the 1960 race where

Nixon lost to John F. Kennedy and the expository plaques get kind of bitchy. In addition to museum-standard fawning praise of Nixon's hard-fought campaign, waged in spite of being hospitalized for a knee injury, there's a recent article from the *Wall Street Journal*'s opinion page speculating on whether Kennedy really won the popular vote after all. It features maps to indicate how close the tally really was. An ersatz living-room set sports a video monitor with footage of the famed 1960 "sweaty Nixon" debate as a narrator explains that while Nixon wore the wrong color suit, refused makeup, and was pale and drawn as a result of being sick, he really won on points and that people listening on the radio thought he was victorious. Besides, the narrator indicates, the debate wasn't really a difference maker anyway. The narrator also said that Nixon could dunk on Kareem Abdul-Jabbar if he felt like it but he just didn't feel like it.

This was followed by a wall of correspondences from children indicating that they supported Nixon and they're sorry he lost. Further displays indicate that some, including Ike, wanted a recount, but Nixon refused in order to keep the country united. The whole "Nixon loses" alcove makes you think of the prom queen somehow not getting the lead in the school musical and getting all crestfallen, bitter, and weepy while the other kids in her class roll their eyes.

More defeat follows, this time for the 1962 California governor's race, and Nixon gets mad again, issuing the famous, "You won't have Dick Nixon to kick around anymore" speech to reporters. Already we had a recurring theme of "Nixon gets pissed off" emerging in the museum, and though it's embarrassing to witness (especially in his own museum), I have to admit it was kind of nice to see. I'm sure that such rage boiling over would not be considered such a great thing if you were an opponent of Nixon, a politically careful supporter of Nixon, or the Peruvian dude he kicked in the shins, but for me, living in an era of carefully packaged candidate/products, I found it refreshing. Instead of the losing candidate being cheerful at the concession speech and saying they're proud of all they accomplished in the doomed campaign, I wish once, just once, someone got up there and yelled, "God-DAMMIT! I'm so much better! Ah, CRAP! You people are STUPID to not vote for me! I'm so PISSED OFF!"

The years 1962–68 are breezed through, and before we know it Nixon's beating Hubert Humphrey (as you or I both probably could) and winning the White House.

I had walked through many rooms of political information so far in this museum and had only now arrived at the presidency. How incredibly different that was from Reagan's displays, which dwelled so heavily on lifesaving, movie acting, snacks, and horses. Then, with a few pictures of campaign buttons, hey, neat, Reagan's president. To Nixon, politics matter. Campaigns matter. Policies matter.

What were these stirrings in my heart? I knew modern conservatism adores, beatifies really, Ronald Reagan. I also knew that I had not yet reached the more notorious bits in the Nixon time line, but I was starting to have . . . feelings for the man. I had to sit down briefly.

Once in office, Nixon had some things to account for, and in the museum, he tries to. Here's the Nixon estate's version of how things went down in Southeast Asia: we bombed Cambodia to cut off Viet Cong supply lines, most people really supported the war but they weren't as noisy as the other people who opposed it, Nixon was the fifth president to deal with Vietnam and his policy there boils down to "I did the best I could." These defenses are accompanied by a little mention of the My Lai Massacre, complete with a photograph of dead children (that I will never forget seeing), and an explanation of how Nixon never stood in the way of the court-martial of a soldier charged with war crimes. Sure, he shouldn't have stood in the way, so there's no need to give him credit for not doing what he shouldn't have done, but the facile lefty depiction of Nixon as a war criminal was not as simple as I might have thought once you look at history. Or at least once you look at the history as Nixon is telling it.

In another display, we are reminded that Nixon also went to China. Got on a plane and flew over there. Shook hands with Mao Tse-tung and Zhou En-lai. No American politician had done that before. People couldn't believe it. Many still can't. He traveled back many times after he left office also (even criticizing the actions in Tiananmen Square at a state event). Because he was there so often, the Soviets reportedly started getting nervous about the United States getting cozy with China, which some say led to the signing of the SALT agreement, limiting the number of strategic ballistic missile launchers. There was more to Nixon than the caricature he had become in later years.

Plus, he went on *Laugh-In*. How cool is that?

The Lincoln Sitting Room is right off the Lincoln Bedroom in the White House. It was Nixon's favorite room in the building and where he wrote his resignation speech. Nixon chose to have this room, not the Oval Office, re-created. I couldn't pay much attention to the room because Nixon's ghost appeared to be in it. In the entryway was a guy in a gray suit with dark, slicked-back hair. He stood with his arms folded and hands tucked in his armpits, his head tilted slightly forward, a bit of a scowl on his face. He was alone except for the docent he was prying for information. In essence, it appeared that Richard Nixon had been resurrected, placed in the body of a good-looking twenty-five-year-old, was roaming the earth, and had chosen to stop by the Lincoln Sitting Room since, as stated earlier, he had always been quite fond of it.

I introduced myself and found that it was not actually

Zombie Nixon but a twenty-five-year-old actor and playwright from Chicago named Nixon Davenport. As nomenclature and current location might indicate, he reeeeeally likes Nixon, and as you might expect, he was pretty charged up to be at the facility. He was on a pilgrimage and this was Mecca.

"Nixon" wore a suit, he said, in order to look better in the numerous photographs he intended to take of himself and, like me, had neglected to shave that morning in order to better capture the Tricky Dick vibe. While we spoke, Davenport, easily six-two, Nixoned himself up, assuming the president's posture, a finger occasionally resting at his mouth. "Nixon" had written two plays about Nixon, including a one-man show based on Nixon's book *Seven Crises*. There was not a great willingness on the part of regional theaters to add dramatic Nixon tributes to their season, so Davenport had been raising money to produce the show himself. Meanwhile, he was about to start rehearsals for a Lee Blessing play he had been cast in about Nixon and Kissinger. He was set to play Nixon. He was excited.

"So you must be a loyal Republican," I said. No, said Nixon Davenport, more of a centrist moderate Democrat. He favored Joe Lieberman in the last campaign.[1] Furthermore, Davenport thinks that if Nixon were around today, he would also be a Democrat in the Lieberman mode.

I asked Davenport if his parents were huge Nixon fans, enough to name their son after the man a full six years after the president stepped down. No, he said, they didn't really have much of an opinion on the matter. His given name is actually Michael, but since there was already an actor in Chicago named Michael Davenport and since he loved Nixon in such a deeply personal way, and since, due to his obsession, he's had the nickname of "Nixon" for years anyway, he decided to use it professionally. Now it's become his name. "Even my parents call me Nixon now," he said. Isn't that every parent's dream?

That the name you give to your child will be jettisoned in favor of that of a president who resigned in disgrace?

As "Nixon" and I talked about Nixon, we drifted around a corner to a room that displayed large panels dedicated to Nixon's (Richard's) policies, the kind of detail that was completely absent at Ronald Reagan's Simi Valley shrine. I explained The Experiment to Davenport and he said that this would be a room I would find especially interesting. He then excused himself to go have some more in-character pictures snapped by seemingly starstruck elderly volunteer docents.

I lingered at the section dedicated to environmental initiatives. As president, Richard Nixon established the Environmental Protection Agency. He signed the Clean Water Act and the Endangered Species Act, a piece of legislation that is practically Naderesque in its far-reaching power. The act essentially mandates the federal government to do whatever is necessary to try to preserve endangered species. While this has led to heated battles over the preservation of natural habitats (like the whole spotted owl thing up here in the Pacific Northwest), it would be political poison for any elected official to try to overturn it. So it stays. Richard Nixon's signature is on it.

In other parts of the room, we learn that under Nixon the voting age was lowered to eighteen and the draft abolished in favor of a volunteer army. While Richard Nixon was president, men walked on the moon. He is also credited with helping establish NASA while serving as vice president and initiating the space shuttle program while president.

Did Nixon himself design the rocket or write all the legislation or single-handedly enact all these measures? Of course not. There were plenty of projects begun way before his term by Johnson and Kennedy, et al., and Nixon certainly opposed some of the things he eventually got credit for. But just as a president bears ultimate responsibility for the things that go wrong under his watch, so too must he earn credit for things that were good. There was a shocking amount of awesomeness going on under Nixon that never gets discussed because we only ever remember the "I'm not a crook" Nixon. As I stood there thinking about Nixon and all his overlooked positive qualities, I thought back to my deceased father's professed fondness for the man. The rest of the family had always chalked up such support to my dad's argumentative and contrary nature in a household filled with liberal and mouthy children, but perhaps there was really something sincere there. Maybe I had misjudged my dad in the same way I had misjudged Nixon. I never got a chance to tell either one of them that they were right before they died. These were unsettling epiphanies to have while standing in the Nixon museum. Who knew that my father, dead nearly six years at the time of The Experiment, would emerge as a recurring character?[2]

Near these displays was a small room of memorabilia featuring the gun given to Nixon by Elvis Presley in that weird photo-op meeting that took place after Elvis had requested to be made a "federal agent-at-large" in the fight against drugs (lots of items with the resulting photo are available in the Nixon gift shop). I got nostalgic for the early seventies, an era I lived through but only as a toddler, where drug-addicted rock stars could meet with presidents and give them guns and everyone's okay with that. What if Snoop Dogg met with President George W. Bush and gave him a Glock as a gift? How weirdly fantastic would that be?

There was a small display of cheerful campaign detritus from the easy 1972 romp over poor George McGovern. It was hard to concentrate on the colorful gear, however, because of the long dark hallway right around the corner. The hallway was the Watergate display.

The position of the Nixon museum was that Nixon had nothing to do with the break-in and really wanted the investigation to go forward without hindrance so that everything could be cleared and the presidency could go on without distraction. Unfortunately, their story goes, there were lots of people out to get Nixon, and when there was a single misunderstood instance of an appearance of an inkling of a shadow of a cover-up, the jackals pounced, clamping down their jaws and not letting go until Nixon heroically left office rather than put the country through further turmoil.[3] Also, presidents had been taping things since Kennedy and the equipment was really lousy, so the famed eighteen-minute gap in the Oval Office recordings, they say, might have been an accident.

Watergate was a mess, we might never know exactly what happened there and most people believe that Nixon had some degree of culpability in the whole affair. Nonetheless, I felt kind of bad for the guy. The man's exhaustive and complex political career came crashing down because, depending on whom you believe, of a stupid mistake, a botched cover-up, or no reason at all. Nixon was not all about Watergate, but his legacy is. We know him today as the "I'm not a crook guy," the "waving before getting on the helicopter" guy, and the guy who gave us the word *Watergate,* which in turn gave us the dreadful suffix *gate* for any kind of political controversy.

The older we get as a society and the more Nixon's presidency fades in active memory, I hope Nixon will be viewed in his full complexity. Once the baby boomers start dropping off, the deeply felt personal attachments about Nixon will fall away and the older establishment of society will be composed of those of us in "Generation X" (we like to put in quotations to demonstrate our detachment from labels while still using those same labels because they're handy). My generation loves guys like Nixon. He'll enter our pantheon of Johnny Cash, Leonard Cohen, Hank Williams, Hunter S. Thompson, and Woody Guthrie as revered icons of generations preceding our own.

Seeing the far end of the Watergate display where Nixon finally gets on that helicopter, you have to wonder what would have happened if there had been no break-in, or if no one found out, or if Nixon had never been implicated. What would he have done in those last two years that were instead occupied by the lifelike Gerald Ford? Would a Democratic outsider like Jimmy Carter have been elected if there had been no beltway disgust for the nation to react against? Would Ronald Reagan have been elected after that in response to Carter? What would the world be like then? Would we have actually ended up in a nuclear war with the Soviet Union that would have wiped all life from the planet? Or would it have been averted long ago? Would there still be a Soviet Union? Would we be in Iraq? Would President George W. Bush be president? Would Dick Cheney's friend in Texas have been shot in the face?

Furthermore, was this complete shift in the course of our history caused by someone on the inside erasing eighteen minutes of Oval Office tape recordings to conceal evidence of a presidential connection to a politically motivated break-in? Or faulty equipment?

The museum wraps up with a re-creation of Nixon's "Eagle's Nest" office in California. There are copies of Nietzsche's *Beyond Good and Evil* as well as books by Aristotle and Machiavelli, biographies of other American presidents, and all of Nixon's own books. Did Reagan ever read Nietzsche? Does President George W. Bush? My head was spinning with crazy Nixon fascination. I was enthralled by the whole complex, environment-friendly, China-going, paranoia-having, space-program-supporting, Nietzsche-reading gestalt of the man.

I had arrived back at the front entrance where the gift shop was. I could not stop comparing and contrasting the Nixon approach and the Reagan approach. Reagan's gift shop featured no books about Nixon, not surprising considering that at no time during the eight years of Reagan's presidency was Nixon welcomed to the White House. Carter had him in, Clinton not only welcomed but consulted with Nixon. Not Reagan. The Reagan gift shop would like you to think that Nixon never happened. There was also nothing in the Reagan shop about Ford, Carter, Clinton, or President George W. Bush. In Nixon's store, however, the presidential world is expansive. There are multiple books about Jimmy Carter; copies of Clinton's autobiography, *My Life;* books dedicated to both Bushes; several selections by some of today's conservative pundits; an occasional tome by historical biographer David McCullough; and, gallantly, a few Reagan books. Also, a book by Larry King for some reason.[4]

I came away from the Nixon museum liking the guy more than I ever thought possible and a little more than I was really comfortable with, to be perfectly honest. Nixon's conservatism said that the government should keep things running smoothly and efficiently, leave you alone wherever possible, and try to do some neat things along the way. I made a note to mail a Nixon biography to Mayor Shawn Larsen of Rexburg, Idaho. Reagan's home planet is a happier place to be, no doubt, but I found Nixon's sullen and complex species-protecting, Mao-meeting, conspiratorial environment to be more like the world I know.

Before leaving, I purchased a Richard Nixon coffee mug, with his smiling face right on it and the top of his head slightly cut off by the mug's brim. It would let me feel like I was drinking his thoughts. I used it nearly every morning for the rest of The Experiment. I saw Nixon Davenport in the gift shop as well. He had just spent $450. If there were one thing I would change about the two museums, it would be to swap personnel. At the Reagan, cheerful and attractive young women had been standing by to help in whatever way possible. They looked like they had been plucked from among the most sophisticated and erudite members of the finest sororities in the nation. At the Nixon, on my way out, I was back to Beverly, still angry at me for thinking I could have received a private tour. "I thought that sounded strange," she railed, rerailed actually since she had first railed several hours before. "We never give private tours. I couldn't believe you asked for a private tour. Nobody gives private tours!" By this point I was sick of taking shit from Beverly, so I leveraged the long-festering Nixon vs. Reagan resentment against her. "Well, the *Reagan* museum was happy to give me a private tour," I said breezily, and walked out.

On the way to the airport to fly home, I gave myself some time off from talk radio and drove in silent contemplation. I felt good about Nixon, but modern conservatives, this group I was attempting to join, did not. Today's conservatives want nothing to do with Nixon. He wanted a big government that shot guys into space and stopped developers from doing things that killed animals and polluted the air and water, but he was conservative in that he wanted to conserve natural resources. He wanted to conserve the nation itself by getting along better with other countries, thus reducing the possibility of war.

Finally, as most people should, he harbored a deep and ever-present fear of hippies.

---

1. So you're the one, I thought.
2. Book clubs: discuss.
3. Between not allowing a recount in 1960 and this, Nixon, like a codependent parent protecting the children, was apparently all about saving the nation from getting emotionally distraught.
4. Maybe their shared love of hunched posture.

## LORD OF THE RINGS:
## THE FELLOWSHIP OF THE RING (2001)

**Summary:** A magical ring imbued with dark powers by the evil lord Sauron has fallen into the hands of hobbit Bilbo Baggins. When he decides to disappear, in part to get away from the ring's power, Bilbo bequeaths the ring to his nephew Frodo, who must then undertake a journey to take the ring to the kingdom of Mordor and toss it into the fires of Mount Doom in order to destroy both the ring and Sauron's power. Joining him is a multispecies coalition of the willing that includes Legolas the elf, Gimli the dwarf, Aragorn and Boromir, both of whom are regular guys, hobbits Merry, Pippin, and Sam, and Ian McKellen, looking like a cross between ZZ Top and Crystal Gayle, playing Gandalf the wizard.

## CONSERVATIVE MESSAGES

- The ring could be viewed as Islamofascism, Sauron as Osama bin Laden, the Orcs as terrorists, and so on and so forth.
- Or the ring could be viewed as communism. Or MoveOn.org. Or Iraq. Or whatever you need it to be given that it's all fantasy anyway and malleable to whatever analogy one cares to draw.
- Frodo could be President George W. Bush, Gandalf could be Reagan. Sam could easily be Tony Blair. Like, super easily.

- You know what? Bilbo could be the first President Bush. Coming into power almost accidentally and then disappearing after a single volume. Maybe if Bush had gotten all bug-eyed and crazy like Bilbo does for the ring, he would have fared better against Bill Clinton and Ross Perot in the '92 debates.
- Aragorn deals with evil ring wraiths by setting them on fire and throwing swords in their faces, which is clearly in defiance of the Geneva conventions but nonetheless effective, because he obviously feels that it's better to fight the ring wraiths on their territory than have to fight them back home.
- The bad-guy Orc fighters aren't really people, deserving of therapy and understanding. They're evil. They're brewed out of mud and then emerge all ugly and angry and ready to kill everyone they find.

## ANTICONSERVATIVE MESSAGES

- Saruman advances a massive deforestation effort, whereas the good guys are more environmentally conscious and would probably oppose efforts to drill for oil in the Shire.
- The fellowship heads only to Mordor, where the actual threat of evil is most present, instead of invading Narnia as part of a larger, more comprehensive plan to eradicate the threat of evil from all fantasy literature.

**OVERALL PERSUASIVENESS SCORE: 51**

# 21

# Courtesy of
# the Red, White,
# and Blue (State)

**In which the author calls upon skills and talents
that he, frankly, barely even possesses to try to
gain a better understanding of the true meaning
of putting a boot in someone's ass.**

**More in the metaphoric sense than literal, of
course. Or so the author hopes.**

My flight had arrived late into Seattle from Los Angeles, and after thinking over the Reagan and Nixon experiences, I was confused and a little troubled. The Experiment was a week from wrapping up, and after all I had been through I still wasn't sure I knew what conservative meant. The goal had been to latch on to this notion of conservatism, then try to buy into it. But what was conservatism? Was it Reagan? Nixon? Lowry? "Gannon"? Kristol? Escalade? Gourley? Rexburg's voting record? Rexburg's mayor? Rexburg's Wal-Mart? José the Nordstrom suit salesman? It's one thing to fail in an experiment but quite another to never get started. At least if you fail, you've made a discovery. If all you do is wonder why nothing connects, I thought, then you're just stupid. Stupid and very very frustrated.

The next morning, all that my beautiful kids wanted was breakfast and maybe a chance to hang out with Dad for a little while before heading off to their respective preschools. What they got, and had been getting more and more of for a while as my personal anxiety mounted, was a moody jerk who could not tolerate any drop of inconvenient behavior. Charlie accidentally knocked over his glass of milk and a flood of anger came rising up from within me. My right eye twitching, I managed to clamp a lid on myself just enough that I didn't scream but not enough to prevent the passive-aggressive sotto voce tirade I dished out about being careful. It was literally spilt milk and I was literally almost crying over it. The kids still loved me, but it was turning into a sad love, the love you have for a pet dog that bites you but it's the only dog you have. The Experiment wasn't cute anymore. Puffy America Shirt was not getting the same laugh as it got a few weeks before at the Wal-Mart where Jill and I picked it out. The sucky singers on the iPod were not funny-sucky anymore. I was confused on the road, unapproachable at home.

Remember how short-tempered Jeff Goldblum got in *The Fly* when the transformation started kicking in? It was like that. Now imagine Jeff Goldblum in *The Fly* trying to get breakfast on the table for his kids and how he'd react to a spatula being thrown across the room. So it was kind of like that.

Something I recall from the acting training I received,[1] besides volumes of psychological abuse and debasement, is the notion of lending the character you're playing your emotions to use in the course of your performance. You may be playing Willy Loman in *Death of a Salesman,* you may call yourself by that character's name, you may say the lines Arthur Miller has written in the script and walk to the places on the stage where the director has told you to go, but it's still your body, your voice, and your emotional life guiding the show that the audience is watching. After you've crafted the character and the performance, you arrive at a much better understanding of what the person you're portraying is going through. I thought it might be interesting to employ a script in my effort to get inside the head of the modern conservative. Perhaps if I fully engaged in the righty thought process, then maybe I would be able to think like them, feel like them, and ultimately be like them. After all, I played Atticus Finch and Anne Frank's father in high school theater and today I have a strong dislike of both racism and Nazis.

So what's the script? God knows there are plenty to choose from, but although many have interesting things to say about the conservative worldview, traditional values, lower taxes, and all that, they lack the kind of substantive emotional life that would make for a good performable text. I mean, you could take a President George W. Bush speech, hone the character, and work on delivering it, but any speech of that sort will have gone through so many speech writers, political consultants, pollsters, and the mind of Karl Rove that by the end it's not much of a piece of drama. You could choose an opinion column by one of the top pundits, but they're either so packed with dispassionate policy discussion or so glib and snarky that there are not many places you can go. I needed something that was personal, conservative, dramatic, and, if at all possible, could lend itself well to public performance. What I needed was my favorite song from the iPod playlist: Toby Keith's "Courtesy of the Red, White, and Blue," and Country Karaoke Night at a country bar called the Little Red Hen.

I called my friend Andy. "Hey, Wednesday night, country karaoke, Little Red Hen," I said.

"I'm there," he replied without hesitation.

I first met Andy when we served tours of duty with the same children's theater in the early nineties. Since then, we've both done quite a bit of acting and then both gotten sick of it.

I became a playwright and he emerged as a director, serving at the helm of three different plays I wrote. He understood the mission.

I had a few hours to go before the big night, so I spent the afternoon rehearsing the number in the comfort of my basement. It was not exactly Olivier's *Hamlet*. Wasn't even Mel Gibson's *Hamlet*. More like Rob Schneider's *Hamlet*. I could occasionally hit the right notes in the song. At other times I could approximate the low-slung country growl that Keith employs so expertly. I was never capable of fully assimilating them into the same performance, but karaoke is karaoke, you're never prepared, so I got in the Bug, wishing I had that Escalade for a better emotional preparation, and drove up there.

The Little Red Hen is an anomaly in Seattle. It's located in the Green Lake neighborhood, where joggers, walkers, bikers, and Rollerbladers, circle a three-mile track around a large pond. Nearby, coffee shops, organic bakeries, upscale bike retailers, and other liberal elite businesses draw in a well-dressed, well-toned, and well-funded clientele, but the Little Red Hen, found right near the massage school (because we somehow need *more* massage therapists in Seattle), is blue-collar all the way. It's a beer joint hosting the kinds of artists who would never think to put the prefix *alt* before the word *country*. I wore Blood Shirt, which loudly proclaimed me to be 100% RED BLOODED AMERICAN, with the Rustler jeans, still-painful cowboy boots, and Flag Hat. "Your best outfit so far," Jill said, marking our most positive encounter with each other in a couple of weeks. I sensed that Jill had arrived at a tipping point with my presence: Was it better to have me around to help take care of the home and family and put up with my petulance? Or was it better that the kids ask questions, the house falls apart, but at least I'm not there stomping around and acting like a total Dick Cheney?

I arrived a little early at the bar to make sure my song was available, ordered a Coors, and grabbed a table. Andy arrived soon after and looked around for me, glancing right past the guy who appeared to have wandered out of a Wal-Mart clearance sale. "Hey! Andy!" I shouted. He gave me a look that made it clear that tonight was not about hanging out with a friend but about helping out a friend who had gone a little crazy. I ordered him a Coors. Within minutes the place started filling up with mostly young folks looking to hook up with other young folks who share their love of Brooks & Dunn, cowboy hats, and having a time that was both "good" and "ol'." The music started, whooping commenced, and I was in one of the only places in the entire city of Seattle that was devoid of irony. People were wearing country gear, enjoying country songs, and nobody was smirking about it. Nobody at the Little Red Hen saw country as something they "discovered." They didn't need to impress their friends with a shocking declaration of what a genius Johnny Cash was or how they've always loved Hank Williams or how Jack White of the White Stripes produced Loretta Lynn's last album so it's okay to like her now. They just liked country music.

The karaoke host called my name. Showtime.

"Courtesy of the Red, White, and Blue" is a tricky song. It starts out as a minimalist ballad. After a quick trip through the basic melody line, a strummed D chord is all that you have to go on for the opening line about American girls. You then get one more strum as you drop down to mention American guys. The rhythm is hard to pick up for anyone on the next couple of lines about the flag, and harder still for me, hanging as I was on the vagaries of a karaoke word display and facing, as I was, a crowded country-western pickup joint.

Soon, much like Styx in "Renegade," Toby Keith picks up the tempo as he talks about his father's eye injury, suffered while serving in the army. He goes on to explain, as did I in mirroring him, how the injured dad wanted Toby, his brother, his sister, and his mother to grow up happy. Wait a second, I thought, he wanted Toby's mother to grow up and live happy? Wasn't the mother already a grown-up? Had Mr. Keith married a child? Shut up and sing, I said.[2] Keith's father was injured during an army training exercise in the fifties and died shortly before the 9/11 attacks. I know this because Keith explains it in a one-and-a-half-minute introduction to the song on his record *Unleashed*.[3] Knowing that context, as an actor, was like having already done the character's back story, and that, combined with feeling the adrenaline of live performance, gave me a little extra oomph as I made my way through the song. Soon I was feeling the character. It was really happening! The young country crowd was getting a little more into it. The girls swayed a little, the fellas nodded their heads to the beat in apparent appreciation of the ass-kicking sentiment that was soon to follow.

Rarely has a song employed so much iconographic and anthropomorphic imagery as "Courtesy of the Red, White, and Blue." Songwriter Toby Keith employs a fist-shaking Statue of Liberty; someone called "Mother Freedom," who has a bell; an eagle flying; and Uncle Sam making a list of places to blow up. But when you sing it, all that imagery, even all those clichés, even the odd ones, especially the odd ones, really help carry it along. There's a nice little rhythm to it and it's easier to sing in a baritone growl when you play the excitement that comes from knowing you're going to blow stuff up and that the Statue of Liberty, much like the Sta-Puft Marshmallow Man in *Ghostbusters,* has come to life and is furiously angry.

Before long, Keith slows it down once again, although he neither backs off the sentiment nor ceases to employ his endless supply of metaphors. He croons about justice being served, dogs growling when their cages are rattled, and the feeling our enemies will have when, upon messing with us, they realize that perhaps messing was inadvisable. Here the song holds for half a moment as the crowd at Keith's live shows and, indeed, the young country hotties at the Little Red Hen, wait, fidget, and hoot a little in anticipation of the big payoff line. The line that got a lot of attention when the song originally came out. The line attacked by liberals, defended by conservatives, and screamed out loud by most country music fans. "We'll put a boot in your ass, it's the American way!" I screwed up the syncopation a bit but still got a big cheer from the crowd, more in support of ass-kicking than of my meager vocal talents, but it still made me feel great going back into the chorus.

As I stood there rocking out, the crowd with me, a group of invisible backup singers coming through the monitors, Andy wondering what had happened to the Birkenstock and Gore-Tex wearer he used to know, it again occurred to me that there's really nothing wrong with the song. Toby Keith was pissed off about the 9/11 attacks and wanted to get the guys responsible. So did I. He also indicated that military retaliation was generally the way the American government responded to such situations, with, euphemistically speaking, rectally inserted footwear. He's right. They do. Then the song was over. "Thenk yew! USA!" I hollered, and returned to my seat.

I was a little dazed about all that had happened and, honestly, I think it was one of the best acting performances I've ever turned in. I had the same feeling I used to get after a really good night onstage where you walk off and don't really know where you are. You're no longer in the world of the play but you have not yet fully assimilated with the rest of the society yet. I had finally been in the moment. Thanks, conservatism!

I managed to drag Andy up onstage for the Toby Keith/ Willie Nelson duet "Beer for My Horses." Just as Willie does on the recording, Andy seemed a bit unsure of himself, but in the Toby role I had built up enough make-believe Oklahoma testosterone to power us both through. The crowd was way into us. They thought we were fantastic. I have a few possible explanations for this, ranked here in increasing order of likelihood, least to most:

1. We were completely awesome.
2. The crowd was filled with aspiring lynch-mob members and we were "riling them up."
3. They were regular people, probably inclined toward conservatism already, who appreciated the song's pro-justice and pro-violence message.
4. They were big fans of Toby Keith, Willie Nelson, or both.
5. It's just a good song, and thanks to leftover musical-theater chops, we did not butcher it.
6. We merely imagined their appreciation.

Regardless, we ordered up another round of Coors and reflected back on our moments of glory as we listened to various singers rock the hits of Tim McGraw, Big & Rich, Reba McEntire, and other country stars. I was again brought back to the notion of simplicity and conservatism. Country singers and country songwriters don't try to reinvent the genre. They don't feel a need to do something bold, innovative, and shocking. There is no country equivalent of Radiohead or Coldplay or Marilyn Manson. Mostly the songs are about regular stuff. A popular hit by Brad Paisley talks about how if your wife or girlfriend cooks all day and the food is terrible, tell her it's great. In exchange, she won't mention the fifteen pounds you need to lose or how you're balding. That's not a lie, the song says, that's love. Hold that up to anything off *Kid A* by Radiohead and you'll see that country music, like conservative politics, is simpler. It's easier to understand. You can get it. I mean no disrespect to Radiohead or John Kerry here either, but people are going to understand Brad Paisley a lot better.

I bade Andy good night. With a successful evening of fake Toby Keith under my belt, I was excited about what could happen once I got to see the real thing. Fortunately, I had already booked a flight to Indianapolis, where Toby Keith would be performing live in concert. I hoped that the night's triumph heralded a new era of mental stability in The Experiment.

---

1. From the same acting teacher as Jeff Goldblum had. Seriously.
2. *Shut Up and Sing* is also the name of a book by conservative talk-show host Laura Ingraham. Do with that information what you will.
3. The intro is actually on the record as its own track and that creates some weird moments on the iPod shuffle option. You'll hear Toby introduce his patriotic fighting song and then suddenly Michael W. Smith is crooning about Jesus.

## LORD OF THE RINGS: THE TWO TOWERS (2002)

**Summary:** The Fellowship has been broken with Sam and Frodo heading for Mordor, Merry and Pippin being captured by the evil Uruk-Hai army of Sauron, and the others taking up with the people of Rohan, a kingdom of humans about to be drawn into the war. Gandalf isn't there because he's thought to be dead. Neither is Boromir because he really is dead. Wars are fought. Gollum, a creepy slimy creature who wants the ring and talks like an asthmatic Carol Channing, is captured and made to guide Frodo and Sam through Mordor. Gandalf shows up again and now looks like a cross between ZZ Top and Edgar Winter.

### CONSERVATIVE MESSAGES

- Like President George W. Bush, all the good guys enjoy running.
- War is something that everyone accepts and prepares for. No one has a "No Mordor War!" or "No Blood for Jewelry!" bumper sticker on their horse.
- "I will not risk open war," says Theoden, king of Rohan. "Open war is upon you, whether you risk it or not," says Aragorn, who is on his way to being king of the world.
- The Ents, big weird creatures that look like trees and walk like retired NFL offensive linemen, don't want to get involved with the battles. "This is not our war,"

they say. "But you're part of this world! Aren't you?" asks Merry.

## ANTICONSERVATIVE MESSAGES

- Sam doesn't trust Gollum as an ally. Frodo thinks such an alliance ultimately has a lot of potential to help accomplish larger goals, so he puts up with the risk. Frodo is a multilateralist.

- Frodo may, in fact, be a full-on liberal. He wants to offer Gollum, who used to be a regular hobbit but was driven mad by the ring, sympathy and understanding. Oh sure, Frodo, blame the parents, blame society or government programs, blame the ring of Sauron.

- It's hard to imagine President George W. Bush being as much of a badass as Aragorn. Aragorn leads armies, battles huge snarling horse-dogs, gets pitched into rivers and almost dies, makes out with Liv Tyler, and unites all species. President George W. Bush takes five-week vacations, dresses up as a fighter pilot, and complains about things being hard.

### OVERALL PERSUASIVENESS SCORE: 42

# 22

### Keepin' It Real

## with **Ford,**

## **Coors,** and

## **Chaw**

**In which the author is someplace where, wow,
the author has certainly never found himself before.
You know when in Superman there was the
Bizarro World where everything was backward
and Superman was a bad guy (urbane liberal readers:
there was a *Seinfeld* episode that dealt with this
as well)? Well, let's just say, Seattle is Metropolis
and then Superman/the author gets on a plane.**

Whereas that first flight to D.C. had felt like a great adventure, this fourth trip, to Indiana, felt like spirit-crushing tedious labor, like I was a traveling salesman with a trunk full of scrub brushes pestering a reluctant customer, except the brushes were conservatism and the reluctant customer was also me. After a long flight, I settled into my suburban Indianapolis motel room. A moth had somehow gotten inside. Normally, I catch moths and put them outside, figuring why kill something that is doing no harm and would rather not be there? But this night, I swatted the thing down with a magazine and then got a cowboy boot out of my bag and bludgeoned the moth over and over until the by now ex-moth became a beige smudge on the motel-room carpet, sure to require plenty of spot cleaning from some underpaid motel maid who will not know what that spot is from. I grunted like the most enraged caveman in the world. What the hell was the matter with me? I've known anger in my day and sometimes honk at other drivers, even cursing them while in my car, but not loud enough for them to hear it.

Finally I slept. Or passed out.

In the morning, I stopped in at Starbucks to get some coffee. Starbucks is headquartered in my hometown and I've seen it grow from a single shop in the Pike Place Market to a worldwide empire of eleventy squizillion stores. Ever since they had dedicated their energies to selling music a few years back, I noticed that most of what they offered was older rock and folk artists. Predictable, nonconfrontational, geared for boomers. But in the past few weeks leading up to The Experiment, they had seemed to shift toward modern singers and bands who were all young, beautiful, earnest, depressed, and completely alien to me. Jill and I had bought one album on the recommendation of a barista, only to discover that it was one of the worst records ever in the history of everything. Fortunately, we were able to swap it out for an old Nat King Cole record they were selling. Ah, Nat King Cole. You old smoothie.

But on this day, or at least at this Starbucks, there was a seeming return to the older artists. The register display was stocked with Etta James, Carole King, and Sly and the Family Stone. I told the barista how glad I was that they had gone back to the nice dependable old music and not this young stuff that I don't understand anymore. On my way out, I noticed a shop next door was having a sale on golf pants. "Oh my goodness, that's a great price on slacks," I thought.

By the day of the Indianapolis Toby Keith concert, I sensed that there had been some movement on the Karl Rove story and his alleged involvement in the outing of Valerie Plame as a CIA operative. The chatter was picking up on the talk radio shows where the righty hosts were declaring it a nonstory but with more strenuousness and desperation than before. The defenses on *National Review* Online's The Corner were getting edgier. The online right-wing message board Free Republic was getting incrementally more hysterical than it normally was. Still, I had no idea what the developments really were and what they meant.

By the time this book hits the streets (unless the liberals give our nation over to China by that point, as talk show host Michael Savage was speculating the night before), Rove will either have resigned, been put in jail, or the whole thing will have blown over. Given the high price paid by Bush administration officials for screwing things up, penalties like Congressional Medals of Freedom and endearing nicknames, I doubted Rove would be inconvenienced in any way.

Maybe that had something to do with what I'd been hearing. Rush told me that the Plame thing is not a story at all, that it doesn't matter a bit, and that's why he's spent the whole week talking about it. He said that there was no way Rove revealed anything to anybody; that Rove may have, in fact, heard about Plame's CIA connection from Robert Novak. Hannity chortled that Democrats are chasing their tails. Limbaugh even pointed to a *New York Times* article, not that I could read it, that he said puts the nail in the coffin of the whole affair.

I talked to Jill the night before on the phone. "Have you heard about this Rove story?" she asked, an urgent gossipy tone in her voice. Well sort of, I said.

"Well, what do you think about it?" she asked. It's at this point in most of our conversations that I would unload a long-winded answer gathered from having read the *New York Times, Washington Post,* the *Wall Street Journal,* a ton of liberal and conservative blogs, and having listened to NPR and possibly BBC coverage. Since Jill's day usually involves more breaking up impromptu preschooler wrestling matches and picking up stray plastic dinosaurs than getting media updates, she relies on me for such information and analysis. Instead I said I didn't know.

"I heard about it, so I picked up the *New York Times*. It's crazy," she said.

"Honey, I . . . uh . . . I gotta go."

Left with mostly conservative opinion in a talk/rant format for my news, I was at least able to take grim delight in how much the talk radio hosts hate one another. This is not uncommon in radio. You have a show named after you, you have people calling in hoping for a chance to talk to you and praise you, and you are rewarded for having strong opinions. Radio folks have egos. But if you feel psychologically abused by them, as I had been feeling, it's nice to hear them stop abusing you and start abusing one another once in a while. Savage called Limbaugh "Hush Bimbo" and Bill O'Reilly "Loofa Boy" (a reference to a sexual harassment charge O'Reilly had recently faced where a female producer claimed that, in a taped phone conversation, he told her how he wanted to scrub her down in the shower). Limbaugh decried the "*Savage*" (emphasis his) and hateful things that some other hosts say. Medved called Savage "an embarrassment to the human race" and implied that Hannity lacks intellectual heft. Hannity never called anyone any names as far as I could tell, but as I tended to doze off during his program, I couldn't know for sure. This infighting was a guilty-pleasure subplot of The Experiment. Like watching a group of crazy fucked-up neighborhood dogs start attacking one another. You know they'll chase you eventually, knock you off your bike, possibly give you rabies, but in that one moment it's a relief to see them tear into one another.

On my way to the Toby Keith concert in suburban Noblesville, I stopped at the store and picked up a large Dasani bottled water (sounds liberal but bottled by the loyally Republican
Coca-Cola Company) and a large bag of my beloved beef jerky,
who always loves me and never judges. I was wearing Puffy
America Shirt, NASCAR cap, Rustler jeans, and the cowboy
boots, which, even if I became 100 percent converted, would
never not hurt. I arrived well in advance of the parking lot being
open for the show, so I headed into the town of Noblesville to
look around. It was one of those suburbs where the architec-
ture is composed almost exclusively of the familiar edifices of
chain stores and restaurants. Wendy's red boxy roof, the slat-
ted roof of Taco Bell, and holding it all together, the expansive
blue roof of Wal-Mart, serving the same central function that
a town square or courthouse might serve in a town that was
actually, you know, real.

After an hour of poking around the Wal-Mart trying to see
if I felt more at home there than I did a few weeks earlier in the
Seattle suburbs, I gave up and left to get something to eat. I
stopped at Chick-fil-A and grabbed a sandwich made of chickens that may have been treated inhumanely because of a lack
of intrusive governmental regulations, a Coca-Cola, and a free
copy of the *Indiana Christian News* newspaper, then I went to
the concert arena.

About half of the Toby Keith fan base looked like regular people you might see walking down the street; the other
half looked like secondary and tertiary characters from a Coen
Brothers film. Further, there was absolutely no chance that
anyone else in the crowd was comparing anything to a Coen
Brothers film. Many of the women wore those small straw
cowboy hats with the front and back parts of the rim pulled
down. Since Toby Keith wears one too, that also makes it cool
for men to wear them, so they did. A lot of the guys really tried

to look like Keith with the wraparound sunglasses, the bent-up hat, the sleeveless shirt, and the beefy body type. One guy had a folded black cowboy hat with the "naked seated long-haired chick" truck-mud-flap chrome emblem. Nice synergy when you think about it.

Not spotted at the concert: anything having to do with the Dixie Chicks. The Texas-based country band had angered many conservatives by saying, while onstage in England, that they were ashamed that President George W. Bush was from Texas. This was seen as treasonous by many on the right and events were held where fans were encouraged to voice their objection, not so much by talking but by bringing Dixie Chicks CDs by a radio station to have them crushed by a bulldozer. Lead singer Natalie Maines had also criticized Keith's song "Courtesy of the Red, White, and Blue (The Angry American)" and Keith responded by knocking her songwriting ability and displaying pictures of her and pictures of Saddam Hussein on a big screen at his concerts. Then Maines responded to that by wearing a T-shirt at the Country Music Awards that read FUTK. If I point out that the TK stood for Toby Keith, you could probably figure out what the FU, although an imperfect acronym, stood for. As previously noted, the Dixie Chicks were not on my playlist.

At the gate, I tried chatting up my fellow concertgoers. By this point, it was raining pretty heavily and a young woman behind me was comparing the weather to conditions at a recent Brad Paisley concert she attended. In my extremely minor capacity as an occasional rock critic, I had reviewed a record by William Shatner where the *Star Trek* actor teamed up with various musicians, one of them Paisley, for a series of duets. I was excited to have something to contribute to the conversation. Maybe it would help me make a friend! "Brad Paisley's on the new William Shatner record, you know!"

"Oh. Really?" She had no idea what I was talking about.

"Yeah, you know, Captain Kirk. It's actually a really inter-esting record. You expect it to be kitschy, and while it is spo-ken word and it is William Shatner doing it, it never really descends into the kind of campy thing you would expect. Ben Folds . . . uh . . . produced it. And Aimee Mann is, um, on there . . . too . . . as well."

Aimee Mann? I was telling a seventeen-year-old Indiana country music fan with glittery eye shadow whose boyfriend had either just returned or was just heading to Iraq about Aimee Mann? Why not deconstruct a Björk album while I was at it? Jesus. I realized that all I was at this point was a weird guy in a silly outfit going to a concert alone. And who goes to concerts alone? Weird guys in silly outfits whom no one wants to talk to. I sank further into my wretched mood.

I picked a spot on the lawn right under a big screen, which showed an advertisement for Coors Light, featuring failed Republican senatorial candidate and beer scion Pete Coors. Hmm, sounded good, so I went and got one.

Where I live, in Seattle, people don't smoke. Maybe it's the close quarters of living in a city or a heightened awareness of health issues, but it simply doesn't happen. I can go weeks at a time without remembering that cigarettes exist. At country music concerts in Indiana, on the other hand, people smoke. At the Verizon Wireless Music Center, there were large mar-keting initiatives dedicated to the joys of tobacco. In the RJ Reynolds Pleasure Lounge (I swear that was the name), you could get two free packs of cigarettes and get to hang out in a cool place where the cool people smoke because it's cool and no one talks about how they're all dying. The catch is that you have to bring in your own pack of cigarettes to show that you are, in fact, a smoker already. I was instantly reminded of the gun-range policy at Wade's where you had to bring in your own gun to demonstrate that either you have a right to own a gun

or sufficient craftiness to steal one. What is it about conservative activities and this system? The backup plan at the range
was to bring a friend. I'm not sure if the Reynolds tent had
such a backup, but since I didn't have a friend the option was
irrelevant. Fortunately, although the smoking people didn't
want me, the spitting people welcomed me with open arms.
In exchange for filling out a registration card and a short sur-
vey about which brands are my favorite, I received four free
cans of, how you say, chaw, including the decidedly nonmacho
peach and berry flavors.

Looking at the crowd, I thought about the difference
between them and me. Differences, plural, actually. They were
with friends, I was alone. They seemed happy, I was miserable.
Most noticeably, they were enthusiastic. Jumping up and down
in their seats, whooping it up long before the concert even
started.

I've never been a whooper. At ball games, I don't do the
Wave. I remain seated, preferring to let the Wave crash over
me. At concerts, I'm more likely to nod appreciatively than
cheer. Often I don't even clap. I am generally the last one to
give a standing ovation, and even then it's more out of not
wanting to be the only jerk not going along with it rather than
having actual initiative to get out of my seat. I mumble through
the Pledge of Allegiance the same way I did in grade school.
Doesn't mean I don't love America, just probably means I'm
Scandinavian, reticent, and kind of a grouch.

But under the auspices of The Experiment, I was trying to
change. There's nothing I could do about my solitude, but I
could change my enthusiasm. Maybe by getting enthusiastic
about this, I could become more of a flag-waving type, a pledge
enunciator, a Yankee Doodle Dandy. Maybe it could all start
with more whooping. I mustered up the same feeling I had

when I did Toby karaoke a couple of nights earlier. Maybe this was that revelatory moment like when people attain Buddhist enlightenment or figure out how to cure their migraines.

So when the first act, Shooter Jennings, Waylon's son, who looks like a cross between his dad and the late comedian Mitch Hedberg, hit the stage, I went bananas. Hooting and hollering like the most-coked up cowboy at the biggest rodeo in Colombia.

"YEEAAAWWGHH, Shooter!" I shrieked, "Awright, Shooter!"

"Shooter!" I added.

Shooter was the opening act and only played for about twenty minutes, but from the way I was acting, this was Hendrix at Monterey. "That's Waylon's kid, right?" I asked the middle-aged guy sitting next to me. Bob Huser, pronounced like the nickname of the state we were in, was a dedicated Waylon fan and was eager to talk about it. He volunteered the information that he has ten favorite Waylon songs because he could never pick just one. He's tried to whittle it down to at least a top five, but it took him forever to whittle down to ten and that's where he gave up. It was apparent that he had spent more than a little time agonizing over this.

Bob had met Shooter Jennings when Shooter was twelve years old. He had also received a personal phone call from Waylon one time after writing a long and heartfelt fan letter explaining why Bob believed, deep down in his heart, that Waylon ought to record a cover of Roberta Flack's "Killing Me Softly." Bob was sure that it would be a number-one hit. Bob loved Waylon in the deepest platonic way a man can love another man and he loved Shooter as well, along with pretty much everything about country music. I had met people like this who felt the same way about R.E.M. or Ani DiFranco, but that same dedication to country was a new one to me. Since Bob was at a Toby Keith concert, I figured he must love Toby as well. He said he liked the music but not the politics.

As for the sentiments expressed in "Courtesy of the Red, White, and Blue," Bob, a Vietnam veteran, did not care for them. "That bastard don't know what the hell he's talking about," he grumbled. "He don't know. Ain't none of these people know what it's really like over there. People are killing, people are dying, and folks here treat it like it's a big party. It ain't a party, it's war."

Bob was a fan of the music but refused to be party to the whole flag-waving enterprise aspect of contemporary country. "This guy [Toby Keith] is six-four, two-forty, let him go over there and fight," Bob said. We spoke most of the way through a set by Lee Ann Womack, a sort of country version of Sarah Jessica Parker, who was second on the bill. Had I access to a private jet, it would have been interesting to fly Bob out to Rexburg to discuss foreign policy with Mayor Shawn Larsen. Had I access to a time machine, it would have been interesting to somehow get Bob Huser in a room with all the various war planners who avoided military service in Vietnam and who composed the American war-planning effort and let them hear what Bob had to say about war. I don't think he would have gone along with the "greeted as liberators" line of thinking. Probably would have thrown a beer can at Wolfowitz. I stood up from the muddying ground, thanked Bob for his insight, and walked through the crowd, looking for a drier spot to watch the concert.

Bob had thrown a wrench into my plans to be an enthusiastic Toby Keith fan. Seems like every time I traveled into what I expected to be the heart of conservatism, I discovered ambivalence. By this point in The Experiment, I actually did kind of like the music, or had at least grown accustomed to it like an imported Moldavian bride in an arranged marriage. I also felt confident that I could feign enthusiasm for at least a sixty-minute set. But Bob had forced me to entertain the possibility that all of this whooping was kind of . . . stupid. Toby was about to hit the stage.

Still, if I acted like I felt the music, then maybe I would feel the music, then maybe I would identify with the sentiment behind it, and then maybe I Would Become A Conservative. It was a long shot, I realized, and my *National Review* contemporaries who are reading this now are thinking, "No, you idiot, that's NOT how you become a conservative!" but time was running out on The Experiment and I needed to try something radical. As it stood, I had become, at best, a mishmash of neo-Nixonian-quasi-Libertarianism with an emphasis on space exploration and clean rodeo and a taste for jerky. No solid foundation, that.

I steeled myself and pledged to go ahead with the whooping. Bob had presented legitimate observations, but I would have to whoop louder, take in Toby, let him into my heart, and learn to live like the happy people in floppy hats lived. One walked by with a cowboy hat he had constructed out of cardboard Coors Light twelve-pack boxes. He was cheered by the crowd for his ingenuity. I joined in.

I knew the show was sponsored by the Ford F Series of trucks, a fleet of huge gas-guzzling pickups designed to make the driver feel like a Big Man. The tickets were very clear on this sponsorship; the F Series mentioned well before Toby's name. But to this point, I hadn't really noticed Ford's presence. Coors Light, Jack Daniel's, Southern Comfort, and RJ Reynolds had made themselves known. We knew what to take to get a buzz, but what were we to drive once we had it? Had Ford sponsored the concert out of the kindness of its heart?

The lights dimmed, we all whooped, and a screen rolled

down. A film began. Science fiction. Aliens patrolling Earth's skies, searching for Toby Keith. Finally, there he is on the screen, driving a Ford F Series pickup, alongside a talking bulldog. The aliens can't catch Toby because the F Series truck goes too fast! They try to rope him, but the F Series truck is too powerful! This went on for, like, ten minutes as the capacity crowd's whooping degenerated into a sort of "guess we'll have to settle for whooping for this" kind of whooping.

Along the way, Toby delivered stilted dialogue that namedropped his popular songs. "They're about to deal with an Angry American," that kind of thing. Finally, the aliens sucked the whole truck aboard their mother ship and tried to shoot Toby with lasers for reasons that were unknown, unexplored, and, if I may lapse into film criticism, likely never considered. Toby told his dog that in case they don't make it, he wants the dog to know that he loves him. "Whatever, girlfriend," said the dog sarcastically. The crowd went crazy with cheering and laughter. Dog! Gay! Hilarious!

Finally, Toby decided enough is enough and to get out of there, so he guns the engine. At that precise moment, a Ford F Series truck crashed through the back wall of the stage and Toby came out with an acoustic guitar with a huge "Ford" logo on it.

Here I was trying to overcome Bob's "that bastard don't know a thing about war" missive and now I had to deal with crass commercialism literally crashing onto the stage at a concert. Other people in the crowd found the presence of Ford highly objectionable, not because Keith is a sellout but because of the make. "Chevy!" they hollered. "Ford sucks!" It was the rare occasion when they had to express it vocally instead of letting the "Calvin peeing on the logo" decal on their truck do the talking. Two songs into the set, Toby ditched the Ford guitar in favor of a plain black one. He then explained to us that they were shooting a commercial during the first two songs and we all get to be in it. Big cheers. And I guess that's why Ford made sure that the ticket prices were low and why we were all given big discounts. Oh, wait. No. Ford got a free set for its commercial, Toby Keith presumably was paid handsomely for his endorsement, and our reward was to be in a commercial and have the honor of helping Ford sell more trucks to other people so they can make more money for themselves and Toby Keith. And all it cost us in the audience was thirty-two dollars for lawn seating or sixty-seven dollars for seating that involved actual chairs. It was all done in the best interests of a large corporation, a muscle-flexing, tough-talking entertainer who is almost a corporation unto himself, and the people of Indiana, a reliably conservative state, went/whooped along with it.

I was doing my earnest best to join them, but the nagging feeling of not being a part of their world would not relent. Then I remembered my new smokeless tobacco that I had been awarded by RJ Reynolds and thought maybe that would help finally get me into the spirit of Toby Keith and the Ford Motor Company at the Verizon Wireless Music Center. I got out a pouch of mint and stuck it between my cheek and gum. Thing is, there's spitting involved with chewing tobacco and I was sitting in a crowd of people with nothing in which to spit. I had my big plastic cup of Coors Light but it was still half full. I temporarily removed my chaw pouch and with two or three big slurps drained the beer.

Oh, I should probably point out at this point that I have never ever tried chewing tobacco before in my life.

I had the pouch in my mouth for about five minutes, dutifully spitting whenever the least amount of liquid formed in my mouth. Here's what I had consumed that day: a chalky scone at the hotel's continental breakfast, a huge coffee from Starbucks, a big thing of Dasani water, a big thing of Coca-Cola, an enormous sack of beef jerky, a Chick-fil-A sandwich, a large and hastily chugged Coors Light, and the first chewing tobacco to ever enter my mouth. It was ten o'clock at night.

Toby started swirling around, I began to lose my balance, my guts felt like they were in the weightless void of space, and I wanted to puke. It had rained earlier in the day and was raining again, heavily, as my chaw high kicked in severely. I sat down on what had become a mudflat. I wanted nothing more than to throw up, figuring at best it would purge this toxin from my system and at worst hasten the death that was surely upon me. I tried to gather myself together while I, a bookish alternative music fan from Seattle, sat in the mud at a Toby Keith concert in Indiana wearing a NASCAR cap and cowboy boots and a shirt that announced that my blood was both American and red. It was as if I was a virus and the body of conservatism was trying to purge me.

Still, had to overcome. Had to persevere. Must whoop. Must overcome Nashville-Industrial Complex. Helping in this effort was the actual music. Unlike moping alterna-rockers who often see their fame, and particularly their big hit songs, as a real hassle, Toby Keith played the hits that people came to hear and they mostly dealt with everyday themes. "Let's Talk About Me" is about Toby getting sick of hearing about his girlfriend's job and life and clothes and he wants to talk about himself (this seems a little disingenuous since nearly every song he plays is already about himself). "How Do You Like Me Now?!" is about the people who didn't like him in high school. "I Love This Bar" is about a bar that Toby Keith loves. They are not imaginative songs, but like the paintings at London's National Portrait Gallery, they are exceptionally well-crafted variations on long-established themes and forms.

As the set went on, there were some surprises. Despite a zealously pro-military position, particularly in regard to anuses and footwear, Keith is more libertarian when it comes to drug policy. He doesn't condemn drug use in general but rather drug use with Willie Nelson in particular during "I'll Never Smoke Weed with Willie Again," a rousing sing-along. Would Nancy Reagan's "Just Say No" antidrug campaign have been any more successful if she used imagery like being curled up in a fetal position on Willie Nelson's tour bus? Would it have been less successful? But it would have been a heck of a lot more interesting, right?

I had expected the country music fans at this show to be enthusiastic war supporters, bonkers for the war effort, blindly supportive of President Bush. While Keith's occasional calls for a hand for "boys and girls over there defending our freedom" was always received enthusiastically, the crowd was more loyal to the notion of partying. A few people, me included, wore America-themed clothing (one guy had a baseball cap that read USA and a T-shirt that read AMERICA, so that whatever you chose to call our nation, he had it covered), but the mood of the crowd was better represented by the fellow in the Coors twelve-pack cowboy hat. Thus, songs like "Ain't Much Fun Since I Stopped Drinkin'" and "A Little Less Talkin' and a Lot More Action" went over bigger than anything patriotic.

Indeed by the time he actually played "Courtesy of the Red, White, and Blue" during the encore, the crowd, Toby Keith, and I all seemed to have lost interest in nationalism. "We'll put a boot in your ass," he sang with about as much passion as the Jiffy Lube guy saying he was going to put a new oil filter in your Hyundai. Keith went through the motions on the song, but his heart wasn't in it. I did my level best to whoop, but I was still a little concerned that any whoop that whooped too hard would result in my whooping up the various beverages, tobacco spit, and beef jerky I had ingested over the course of the day. The crowd cheered a bit, but they had been stuck in the mud, in the quagmire as it were, for what had turned out to be of much longer duration than any of us were expecting. Finally, after a long slog, Toby Keith withdrew his musical troops from the situation and bade us all a casual good night.

Due to poor planning on the part of the controlling authorities, I was mired in the parking lot with a bunch of other motorists and we had no idea how long we'd have to remain in that inhospitable environment or when we'd ever see our families again. After breaking free of the lot, forty-five minutes later, I got lost on obscure unlit rural Indiana highways. It was twelve-thirty in the morning. I called Jill on the cell phone. "Where are you?" she asked.

"I don't know," I said. "What part of Indiana is it when you see a farm and a truck?"

"All of it."

"I think I'm going to need to live here now. I don't think there's any way I am going to get back to the freeway. You can come live with me if you think I can ever be found."

Finally, completely by accident, I stumbled across I-69 and made my way back to Indianapolis at one-thirty.

Being rained on for a few hours might have had something to do with this, as might what could only charitably be described as my diet, but looking at myself in the mirror at the hotel that night/morning, I was struck by what a wretch I had become. The little flecks of gray that used to show up intermittently on my sideburns before The Experiment began had made large colonies now and were threatening my entire head like that black goo on *The X-Files*. I was farting almost constantly and my hair, parted neatly to the right when this all got started, was now a swirling mass, a whirlwind on the top of my head. It looked like Doppler radar of an approaching hurricane. I was ugly on the inside as well. I had stopped making friendly chitchat with cashiers, the twitch in my right eye was getting worse, and I scowled all the time.

Dropping off the rental car at the airport the next morning, I was forced to interact with a perky and incredibly attractive young woman who was working the counter. She tried to make small talk with me, asking how my visit had gone and commenting on how cute my Apple laptop was. I barely responded at all. I scowled. She asked what I used the laptop for and I told her, gruffly, that I was writing a book. "Wow! That's so fascinating! I'd love to hear more about it!" Okay, you have to understand this because it's very important: for funny-looking, heavily married guys in their midthirties, this kind of thing happens about as often as Democrats winning the White House. It's like finding a hundred-dollar bill in an old coat. Opportunities to flirt with pretty young women in the customer service industry are completely harmless but let you feel you still "got it" without the threat of anything actually happening, because you know you're married, she knows you're funny-looking and married—she's just being friendly is all or maybe merely charitable—and in a few moments, you'll leave each other's company forever. Guys like us revere moments like this and sometimes discuss them with a gentle fondness with other guys like us. So what did I do when she said she wanted to hear more about the book? I sneered "well, it's not done yet" and went out to sit and stare straight ahead in the empty shuttle van. Her perkiness, prettiness, and innocence were no match for my mental fatigue and self-loathing. It may be ten years before I ever get that chance again.

My flight was at 7 A.M. but I had arrived two hours early. I wanted to leave Indiana. Nothing against the fine people of the state, one of whom had been nice to me right there at the end, but overall I had not had a pleasant trip. The Toby Keith con-

cert (and again, Jonah, Rich, and Bill, I'm sorry I went looking for political meaning at a Toby Keith concert) had taught
me only that Toby Keith is a phony. Not that there's anything
necessarily wrong with being a big phony, but when you're a
phony while trafficking in the arenas of war and politics, that's
dangerous.

I was glad to be on an airplane again because it was my last
flight of The Experiment. Soon I would be home. I wasn't sure
my family would like me all that much, but home was home
and at least I would have more golf shirts to choose from for
the remaining three days of The Experiment. I was still toting around my now tattered but still unfinished copy of Sean
Hannity's *Deliver Us from Evil,* but I didn't read. Reading
meant thinking. I couldn't think anymore. I put on the iPod,
selected the shuffle option, and leaned back.

Cruel, cruel iPod picked Toby Keith. In "Rock You Baby,"
Toby sings about how he's planning to "rock" someone who's
sad. Although *rock* as a verb has multiple definitions, I think he
means he's going to cradle that person and gently sway them
back and forth. This is different, of course, from what Jon Bon
Jovi speaks of in "Wanted Dead or Alive," where he boasts of
seeing a million faces and rocking them all. Sadly, in that song,
Bon Jovi never explicates precisely how one goes about rocking a face, let alone a million of them. Jon Bon Jovi is a big
contributor to the Democratic Party and hosted fund-raising
concerts for John Kerry in 2004. Meanwhile, Bon Jovi guitarist
Richie Sambora (you always have to wonder if he's ever been
resentful of spending his life in a band named after another
dude) married actress Heather Locklear, a known Republican.
Sambora also used to date Cher, a liberal critic of President
George W. Bush, who was formerly married to Sonny Bono,
who was elected to Congress as a Republican. Locklear was
once married to Rocker Tommy Lee, whose political affiliation

is unknown at press time but is something of a drug-addled sex fiend. Politics and show business are tricky affairs. This was typical of the chains of thoughts I was encountering at this stage in The Experiment. Everything was politics, everything was ideology, everyone was separated into camps. Everyone was at war. Reds vs. blues, right vs. left, R vs. D, Susan Sarandon vs. that lady who plays the mom in *Everybody Loves Raymond* and is really pro-life. My brain was always on and always tuned into an episode of *Crossfire.* Worse yet, it was episodes of *Crossfire* that featured Bon Jovi and Cher, which is not a pleasant thing for your brain to do to you.

Anyway, in "Rock You Baby," Toby Keith promises to rock a "shattered lady" in his arms. He pledges to shake her emotions right down to her soul, love her all over right out of control, and make her crazy over him. Are such overwhelming experiences really what a shattered lady needs? Again, I returned to Bob Huser and his comments about Toby Keith not knowing what he was talking about when it came to singing about war. I don't think Toby Keith knows what he's talking about in regard to shattered lady counseling either.

## LORD OF THE RINGS:
## THE RETURN OF THE KING (2003)

**Summary:** In the final installment of the trilogy, Sam and Frodo get closer and closer to Mount Doom and their goal of destroying the ring. Gollum helps their cause while also undermining their friendship, sowing mistrust between all parties as part of his plan to get the ring for himself. Lots more battles are fought involving all sorts of people whose names are all goofy-sounding as Sauron continues his goal of world domination. In the end, and although it nearly kills him in the process, Frodo tosses the ring into the fires of Mount Doom, thus saving the world. The hobbits return to the shire, Aragorn becomes king of the world, and everything is more or less fine.

## CONSERVATIVE MESSAGES

- Wars can be won if everyone just stands firm.
- Sometimes the best way to accomplish your objective is to leave your allies behind and head off on your own to do what you think is right.
- Bad guy Sauron is all about big government. He tries to control all of Middle Earth, he conscripts massive armies, and he spies on everything he can. One imagines that he probably taxes the hell out of everyone and operates a massive and wasteful socialized medicine system. Meanwhile, the fellowship is all about

states' rights, with each kingdom having the power to decide what's right for their own citizens.

## ANTICONSERVATIVE MESSAGES

- Legolas and Gimli enlist an army of the dead to help the fellowship defeat Sauron, implicitly endorsing something approximating witchcraft.

### OVERALL PERSUASIVENESS SCORE: 70

# 23

## What

# Ended Up

### Happening

**In which the author arrives at the finish line and is
whisked off to an important policy-adviser position in
the President George W. Bush administration or moves
into a bungalow behind Sean Penn's house to publish
leftist pamphlets or maybe something else happens.**

Back from Indianapolis, I had no more trips planned, no
more test-driving of SUVs or days at the gun range, and
with no clear directive, I was adrift. On the planning calendar, these last two days were marked, simply, "Reflection." It
seemed so idyllic at the time I penciled that in.

One bit of good news involved my volatile and angry
demeanor. No longer was I engaged in the type of behavior
that prompted my son, my beloved boy, to say, "Maybe when
you want me to do something, you should use nice words,
Dad. Instead of being mad, you should say nice things and that
would be better." I make a lot of jokes, but to hear a sweet and
wonderful four-year-old kid say that smashes your heart. Yet
instead of snapping at the kids and making life yet harder for my
long-suffering and still inexplicably faithful wife, I had become
docile. As if neutered. No longer a raving jerk, I was a socially
acceptable husk of my former self: quiet, defeated, and pliant.

In fact, I'm pretty sure I had gone what might be described as somewhat "crazy," but drifting hollow-eyed through society is more polite than exploding with unprocessed rage.

"Don't worry, it will all be over soon," Jill said as The Experiment drew to an end. I couldn't be 100 percent certain it was The Experiment she was talking about. I also didn't know how I was defining my politics anymore. There were plenty of liberal parts left over that I had not yet abandoned. Likewise, I had found some attractive points on the right, but they had not gelled into a worldview.

So I had scraps and fragments of a thousand different ideas all swirling around, mixed with a hatred of humankind and a joyless beef-jerky dependency. I figured if I had skimped on anything during the four weeks of attempted self-persuasion, outside of sensible wardrobe decisions, it was scholarship. I had been so busy with the guns and the Toby Keith and the finding out what was so bad about The Gay that I hadn't bothered to read a whole lot of books from serious people with well-thought-out ideas about conservatism's actual merits. If I was going to give conservatism every chance I could, I might as well shove a few more books into my head and wait, now with nervous urgency as the shot clock wound down, for the Zen-like moment of enlightenment.

But what to study? My Nixon man-crush, and the resulting conservative identification that went with it, had been okay, but I still didn't feel attached to *neo*conservatism, the philosophy that is said to dominate the Bush administration and get us into, you know, wars and stuff. So at the Admiral district branch of the Seattle Public Library near my house, I settled down to read Robert Kagan's *Of Paradise and Power: America and Europe in the New World Order*. Kagan's name is one of the first to come up in any discussion of neoconservatism since he, along with Bill Kristol, was a founder of the Project for a New American Century, the leading neoconservatism think tank. I had found some things to like about neoconservatism early on in The Experiment—its allegiance to the New Deal, its fealty to civil rights, Bill Kristol's eerie resemblance to my late father. But it was a hard line of thinking to hang with in the summer of 2005 when every day brought reports of new insurgent attacks in Iraq targeted at our strong and aggressive stool-leg #3 military. The world seemed to be saying that no, in fact, it did not particularly want to be saved and here's a car bomb to illustrate that point. In that context, it was hard to hang with anything Kagan had to say. He argued the European way of doing things was outmoded since they tend to accept imperfect solutions to their problems and favor diplomacy over military solutions. This, he argues, is because for a lot of historical reasons, they're relatively powerless. So they band together to try to create some semblance of communal power based on whatever remains of their former glory, much like the Traveling Wilburys. Except instead of music videos, they make the euro, and because they rely on the more powerful United States for emergency military protection, they can spend all their money on their own economic prosperity.

I tried to be critical of the silly Europeans and their whole "reluctance to embrace military solutions" thing, but then I remembered the news report I had just heard, during Rush Limbaugh I think it was, about twelve people being killed that day in various insurgent attacks as well as a car bomb the previous weekend in Musayyib that had killed at least seventy-one. Just as my collegiate communist tendencies had been trumped by the idea of having to pay bills, so too was neoconservative military policy being trumped by bombs.

Then someone shoved a drill into my forehead. At least that's what it felt like. It was the worst headache I ever had and it came on in a flash. "Am I having an aneurysm?" I asked.[1] Earlier on in this whole undertaking, I may have pondered whether it was my body fighting off the infection of conservatism or the final pains of liberal detox before I opened up into the clean and sober world of the right. On these, the final days of The Experiment, however, I didn't care to draw such comical suppositions. Lee Greenwood, when you listen to his music, starts out as just some singer, then he becomes comically sucky, then he crosses the threshold of actual, verifiable, measurable pain. That's what The Experiment had become. It hurt. Physically and psychically. As I read Kagan's book in the study carrel, it all got to be too much, I began to shut down, the words began to slide off the page, and I started to doze off, hobo style, right in the middle of the damn library. That guy at Nordstrom had a pretty good point: What *was* the meaning of life?

Academic conservatism being unproductive, I opted to give cultural conservatism one last shot and take Charlie to a base-ball game at the home of the minor league Tacoma Rainiers. Maybe a trip to the ol' ballpark with my dear son would finally kick me into gear, affect a Norman Rockwell vibe, and every-thing would click. Perhaps some quality time, I desperately posited, will mitigate the therapy bill he sends me years from now in response to all that had happened in the last month. I put on the NASCAR cap along with a light blue golf shirt, khaki shorts, and Top-Siders. Disjointed, yes. As was I.

Unfortunately, my card-carrying Sierra Club member off-spring was fixated not on baseball but on environmental issues in the Arctic National Wildlife Refuge as we sat in the stands watching the game, he with his cotton candy, me with an all-American hot dog.

"Dad?"

"Yeah, Charlie?"

"Um. Do you know who Bush is?"

"President George W. Bush, you mean?"

"Dad, Bush wants to kill the polar bears and the arctic foxes so he can drill for oil! We need to stop him and make him not do that! We need to saaaaaave the animals! DAD! SAVE THE ANIMALS!" A few people in the stands craned around to look at us.

"Well, I don't think President George W. Bush personally wants to kill the animals."

"Oh. Well." Pause. "I think he does. I think he wants to grab the animals in his hands . . . and KILL them!"

"Is that what they've been saying in your Sierra Club newsletter?"

"No. It's just what I think. Dad, when Bush plays with someone he hurts them. He wrestles and hurts their bodies!"

"Really? He does?"

"Yeah. He should use his words, Dad. DAD! We need to make Bush use his WORDS!"

I tried to explain how baseball works since he was still unclear on which players were on each other's team and why it makes a difference who wins. Charlie was still more prone to simply root for excitement itself, cheering loudly whenever anyone got a hit or whenever the team's reindeer mascot came running out or just when someone else was cheering loudly. He loved all that, but Charlie seemed distracted by the Arctic National Wildlife Refuge throughout the game, which made our conversation stilted. "Do they have reindeer in the Arctic where Bush wants to drill, Dad?" he asked, watching the friendly anthropomorphized reindeer gallivant about on the field. I told him I didn't know. He said no more on the subject but gave me a look that said one thing: support Bush and you're killing that reindeer. Dad.

John McCain was on *The Tonight Show* late that night. I stayed up to watch, hoping again for epiphany, but all he talked about was his part in the movie *Wedding Crashers*. It didn't strike me as odd that a conservative Republican was promoting his role in a movie that featured abundant nudity and debauchery. Nothing seemed odd to me anymore. After weeks of country music and pundits and Idaho and Nixon, I felt about as ready to become a conservative as I did to remain a liberal or possibly join a reconstituted Whig Party.

The next morning I sat at the table for breakfast shoving a soggy pile of Raisin Bran around a bowl. Jill and the kids got ready for a summer day of fun and outdoor play, maybe a trip to the splash pool at the government-funded city park. I figured to go back to the library and give inspiration, maybe in the form of an Irving Kristol treatise, one more chance to strike.

It was the final day of The Experiment.

How could mere conservatism and country music have made me such a wretch? The people to whom this philosophy was native seemed like happy folks, like in Rexburg, Idaho, where Bush gets 92 percent of the vote and everyone gives birth to two or three babies before lunch every day. Or like the glib confidence expressed by my alternate-universe friends at *The National Review*. Why had conservatism served them so well and torn me to shreds? These things are a bummer to contemplate while sitting at the breakfast table with your small, perfectly innocent kids, so I got dressed (Puffy America Shirt, of which I had grown oddly fond, would be the T-shirt of The Experiment's final day), said good-bye to the family, and headed for the library.

"Okay, great," I mumbled to myself driving along the Alaskan Way Viaduct through downtown Seattle, "turn my world upside down, travel the country, drop a grand on a new suit, subject myself to lethal amounts of Michael W. Smith, and what's the finding? What is the great knowledge gained?

Where did I end up? Grouchy and disoriented. Well, that's fucking perfect. Way to go, Magellan." Thinking about these issues, I got distracted and missed my exit. Instead of turning onto Seneca Street, which is how I would normally get to the library, I kept going straight. The only other downtown exit was completely congested, so I elected to keep going through the Battery Street tunnel, putting me north of downtown, figuring I'd then loop back downtown to the library. I turned right at the old Hostess snack-cake plant on Aurora Avenue and into this odd Seattle neighborhood just south of Lake Union that people usually call "South Lake Union" rather than try to figure the place out sufficiently to give it a proper name. South Lake Union was once an industrial section of the city, but as the city has become less industrial, the neighborhood has become more or less a conglomeration of enormous low, flat buildings that puzzle people.

There had been talk years ago of leveling the aging warehouses in this part of town and turning it all into a massive park. As is generally the case in Seattle, many meetings were held, studies were conducted, multiple votes took place, computer renderings were made, lots of people got very emotional, and, in the end, not a damn thing got built. In recent years, South Lake Union was targeted to be a biotech hub. That may yet happen, although as I drove through it that day, it still looked much more like the abandoned remnants of a once-thriving meatpacking industry.

I get to this part of town once or twice a year, tops, but it always fascinates me because everybody wants it to be something grand but no one can ever get a plan together that makes the neighborhood anything in particular. I had always felt fond of the area for precisely that reason. I intended to turn south and head back downtown for another crack at the library, but something made me not do it. Instead, I drove around South

Lake Union for a while. There were some warehouses that had been turned into performance spaces, some that had become imported-furniture showrooms, and some that were still warehouses. There were ancient delivery trucks next to late-model Land Rovers. Architect offices, a basketball court, a deli, and entire blocks that looked pretty much empty. It had never become a park, it was no longer an industrial zone, it wasn't a biotech hub. It was just this part of town that I ended up in accidentally while engaging in too much self-loathing to catch my exit. There was something intriguing and highly resonant about it.

I stopped the car and got out. I looked around. Then I thought. Looked some more, thought some more. Then I just stood there, mouth slightly agape, dressed in Puffy America Shirt, golf shorts, and Top-Siders. I may have been like that for a minute or three hours. Gears were moving in my head.

"That's it!" I finally said out loud. This part of Seattle, this "everything and nothing all at once" neighborhood, was ME. My brain! A month ago, I had set out on what I thought was a mental journey from liberalism to conservatism. I never arrived at the destination of conservatism, never got to a point where my brain believed that all political and social beliefs could be conveniently filed under that one word or put neatly atop the three stool legs. By the same token, I had not begun the journey from a point of absolute liberalism. Sure, I grew up in a house of Reagan-protesting Europeans and went on to live in a city that is as left politically as it is geographically on a map, but I had long chafed at the liberal label as being too facile. So in terms of The Experiment, I had thought I was taking a trip from, say, Los Angeles to New York, when in fact I was actually beginning in, like, Provo, Utah, and ending up in Tampa. Or Belgium. Or space. Or South Lake Union. "This place is my brain!" I hollered breathily.[2] I was onto something, I thought, experiencing a surprising surge of positive emotion. I confidently strode around South Lake Union, the neighborhood that stubbornly refused to be labeled as anything in particular.

Here I had been sinking into a funk because I had neither arrived at rock-solid, Buckley-worshiping, gun-toting, The Gay-fearing conservatism nor rejected conservatism entirely, bringing me to a place where I knew for an absolute fact that the left was right and the right was wrong. I was despondent, thinking The Experiment had been a terrible failure. I was in turmoil after blasting my brain full of all sorts of thoughts, ideas, sounds, and images that don't generally make it in there.

It's like having houseguests, tons of them, and they're all armed, fond of Charlie Daniels, and they like to eat a lot of meat. When you pack a month's worth of those kinds of guests into your Seattle home, conflicts will occur, just as they would if someone from some conservative town in the Deep South tried the inverse of The Experiment.[3] From there, you can choose your response. Most folks in Seattle or any other liberal community for that matter choose to kick all those guests out, then bar the door, change the locks, turn out the lights, hide under the furniture, and never let those guests in again. Makes everything easier, less stuff to clean up. Another option would be to let everyone in the house fight it out, like on that *Big Brother* show on CBS, and try to decide who's right and who's not, declare a victor, and then pledge undying loyalty to either the right or the left. The most interesting alternative, however, is to just accept that it's okay for your head to be a crazy conflicted noisy party all the time and try to keep the conversation lively and informative.

The free-trade people make some pretty good philosophical points about freedom, as do the folks who point out the negative repercussions of that same policy. You can close your ears to whichever is different from what you've always thought, you can choose one or the other as Your Belief, or you can just say, "Wow. Good points," and embrace that complexity. The world is more interesting when you experience it with open ears. You can hear a lot of things.

When I was at the College Republicans convention a few weeks earlier, I had wondered how conservative, on a scale of 1 to 100, I would be by the end of The Experiment. But it's not that linear. I had arrived at somewhere outside the quantitative valuation, somewhere not measured in those simple terms. The scale didn't apply. I wasn't a 1 or a 45 or a 92. I was, like, an elk, or the Eiffel Tower, or the Pythagorean theorem.

Much is made in America of the war between the left and the right. Supposedly, we are divided between liberal and conservative, blue state and red state, and we all hate one another. But two of the most interesting people I met over the course of The Experiment stood outside that strict binary proposition: Mayor Shawn Larsen of Rexburg, Idaho, with his government-friendly, devoutly religious, anti-Bush views, and Nixon. The least interesting people were the ones who seemed to start with identifying themselves as "conservative" and then tackled the issue with those blinders on. I've met plenty of uninteresting liberals who do the same thing.

Maybe if you shut your mouth and open your ears, you'll hear something that makes a shocking amount of sense because it's coming from someone who really believes it instead of coming from someone belittling that person's beliefs. Even if you don't come away from that listening session in agreement with what that person's saying, at least you can get it. Let a liberal explain socialized medicine. Let a conservative explain tax cuts.

By the last day of The Experiment, I had seen some of my beliefs tweaked, some left unchanged, and some where I understood many different positions and simply didn't know what was right and wrong or if there even was such a thing as absolute right or absolute wrong. More than anything, I had had my brain stretched out a few sizes. I hadn't proven that you can switch from liberal to conservative through massive doses of righty rhetoric and culture, but I had proven that you can blast yourself out of your comfort zone and get, if not a smarter brain, at least a wider one. Curiously, once I realized this image of a wider brain, my pounding headache seemed to lift, and within a few hours, the eye twitch faded away as well.[4]

After my long and thoughtful walk, I stopped in at a coffee shop and took an inventory of my opinions. They mostly ended up sounding like common sense as opposed to orthodoxy. Lower taxes are great but only if the government is not spending too much money. Moving further around the stool to leg #2, I've already been living a version of traditional values for a while. My kids say please and thank you. We go to church. I go to work while Jill is at home with the children. She cooks. I own The Suit. Our kids will hopefully help take care of us when we're doddering fogies, thus unburdening the state from having to do so. We've done all these things because they seem like the right things to do. Never did figure out how The Gay posed a threat to all that. Oh well.

As for stool leg #3, the role of the military, I thought about what Illinois senator Barack Obama said about an antiwar protest: "I noticed that a lot of people at that rally were wearing buttons saying, 'War Is Not an Option.' And I thought, I don't agree with that. Sometimes war is an option. The Civil War was worth fighting. World War Two. So I got up and said that, among other things." What he said, among other things, was "I am not opposed to all wars. I'm opposed to dumb wars." Invading and occupying Iraq, he said, would be "a rash war, a war based not on reason but on passion, not on principle but on politics." Obama is a Democrat, but what he was saying was conservative.

In other matters:

Beef jerky: Pro!

Lee Greenwood: Con!

Guns: Be careful with them and bring Larry with you if at all possible.

SUVs: Darkly appealing.

Country music: Like reality television, it's not exceptionally mind-bending entertainment but entertaining and effective when done well. When it's done poorly, everyone suffers.

NASCAR: I liked the cap I bought but still didn't get the appeal of watching several fast cars constantly turning left.

After immersing myself in that way of thinking for a month, I had come to understand conservatism in a deeper and richer way than any other liberal I know. I met a ton of nice people with sincere beliefs that they felt were crucial to making America a better place. I saw no evidence of evil neoconservatives drinking the blood of innocent civilians while plotting ways to enrich multinational corporations. Mostly I saw people doing what they thought best.

There were six hours remaining in The Experiment. As I drove home to see whatever remained of my family, I heard a name on the radio that I had never heard before: John Roberts. He had just been nominated for the seat on the Supreme Court vacated by Sandra Day O'Connor. He'd go on to replace Chief Justice William Rehnquist when Rehnquist passed away, promoted even before being hired. On this, the last day of The Experiment, people didn't know a lot about Roberts yet. He may have been a conservative, but there seemed to be a lot of ambiguity in his record and no clear indication of how he would vote on any particular issue once confirmed to the bench. Most people were taking a wait-and-see approach to him. He defied easy labeling.

That night, I changed out of Puffy America Shirt and into a handsome blue polo shirt and khaki slacks. "Are you done thinking about things for the book, Dad?" Charlie asked me. I told him yeah, kind of, pretty much. But also not at all.

"Do you love Bush now?" he asked gravely.

"I don't think I love him. But I don't hate him either. He's just doing what he thinks he should do."

"Do you love the animals in the Arctic?"

"Yes. I love those animals," I said. That seemed to satisfy him completely. I made a mental note to see if there were any preschool drum circles I could sign him up for.

That night, as The Experiment ebbed away, Jill and I got a sitter for the kids, went out to dinner, and had a lovely time. She said the kids really missed me and were looking forward to our upcoming vacation in Montana together (nowhere better to unwind from a month of conservatism than the home state of the Unabomber and various paramilitary militia groups). She said she was glad that she wouldn't be seeing me in the Wal-Mart clothes anymore. She asked if I was glad to have The Experiment over. I told her yeah, I wanted to expand my musical tastes again and I was looking forward to listening to *Morning Edition* on NPR the next day. "But those polo shirts really look good on you, John," she said. "You should keep wearing them."

"Thanks," I said, "I think I will. Some of them actually fit pretty well."

---

1. I instantly mourned my own death but also was a little jazzed about such an unexpected dramatic ending for The Experiment.

2. Helpful revelations were coming, but I was still unstable enough to declare things like that out loud in public.

3. Which I desperately hope they do, and write to me to tell me how it went.

4. Knowing I was hours away from purging Clint Black from my musical life probably helped that along.

# acknowledgments

In a general sense, I would like to thank all the people and events from my entire life.

Now, on to specificity. My brilliant and generous agent, Jennifer Gates of Zachary Shuster Harmsworth, plucked me from obscurity and with almost eerie amounts of patience and encouragement, guided this thing through. I'll never be able to thank her or those at her agency enough. Mauro DiPreta at William Morrow understood this book before most anyone else, including me, and thanks to him, Joelle Yudin, Ben Bruton, Lynn Grady, and everyone at William Morrow and HarperCollins.

Much gratitude to my good friend the brilliant Bill Radke, who would have been an inspiration even if I had never met him and if he hadn't given me so many opportunities, which he totally did. Along those lines, thanks to Wendy Sykes, Derek Wang, Dave Snyder, Luke Burbank, and Jeannie Yandel. And along *those* lines, thanks to Jeff Hansen, Dave Beck, Cathy Duchamp, Phyllis Fletcher, Arvid Hokanson, Ross Reynolds, Steve Scher, Marcie Sillman, Megan Sukys, and everyone at KUOW.

There are a whole lot of people I've known who not only believed in my various visions and projects but who in many cases went to a lot of trouble to make them happen. So thanks to Dave Liljengren, Susan Benson, Annie Howell, Chris Shainin, Craig Gunsul, Jennifer Holman, Doug Hunt, Anthony Winkler, Kirk Anderson, Sam Goldberg, Heidi Robinson, Chris Burns, Tom May, Heidi Pickman, Krissy Clark, everyone at *Weekend America* in fact, plus Sara Sarasohn, and Gary Waliek.

Several years ago, I sent in a humble, unsolicited submission to the McSweeney's Web site, the editor of which opted to print it. Thus began my association with an organization I deeply admire and am humbled to be a small part of. Thanks to Dave Eggers for his generosity and for the important and incredibly cool work he does. Much love also to John Warner, Kevin Shay, and the entire McSweeney's Army, which might not defeat the Kiss Army in a war but still, you'd have to admit that would be a fascinating war to watch.

Reaching further back, thanks to Heather Evans for her timely advice, Dana Burgess, Merlin Epp, and the Nickels family. I've been blessed with people in my life who are both dear friends and wise counselors and I thank Sean Farnand, Hunt Holman, and Andy Jensen for being who they are. Sincere thanks go to the members of Free Range Chickens and Chicken Starship, two rock bands I've been proud to perpetuate along with Joe Chicken, Sean Chicken, Scott Chicken, Steve Chicken, and all the Bass Playing Chickens we have known.

Thanks to everyone who helped in so many ways during the process of writing this book, including Larry and Beth Olson, Karen Kassameyer and John LaPlante, Elizabeth Prescott, Misti Covitz, John and Frances Smersh, Becky Dobbins, Charles Kiblinger, Heather Dahl, Nancy Pearl, Henry Alford, Glenn Fleishmann, Matthew Baldwin, Kevin Guilfoile, John Hodgman, Juliet at Uptown Espresso, Michael Cohen, Kimm Viebrock, and the music of the Postal Service. Deep thanks and love to my family for a lifetime of inspiration and support; thank you Mom,

Dad, Mette-Line and Thyge, Rick, Mark, and Lisbet (a trusted writing associate who skillfully doubles as my sister). A blessing of my marriage is that I get to have another fantastic family, so laurels of gratefulness go out to Susie, Molly, DJ, Jenny, Peter, Jinsey, Vernon, Billy, and all my zillions of nieces and nephews on both sides.

Finally, the three people who I can never begin to thank enough, listed in ascending order of how long I've known them. My wonderful daughter, Kate, has provided unwavering love, joy, and inspiration. My darling son, Charlie, is my muse, my friend, and a source of constant wonder. And above all, I thank my wife, Jill Moe. She's my friend, my sharpest critic, a source of strength, a font of wisdom; also she's hot, and without her constant and massive support, love, and dedication, there would be no book. Thanks, sweetie, I love you.